FROM THE DEPTHS

Jeffry Weiss
3595 Hayden Pl. Suite #1
Boulder, Colo. 80301
(303) 525-1771
jeffryweiss@gmail.com

1

FADE IN:

MONTAGE:

SCENES OF POLLUTION from all over the world: open sewage spilling into the oceans in Brazil, The Philippines, Indonesia, Spain, Italy, Greece, Africa and America.

The coral reefs turning white, crumbling and dying. Every type of fish, small to large, predator and prey, lying dead or dying in shallow waters or washed up on shore. Surfers and swimmers get out of the water with large red welts all over their bodies.

Boats near shore and out at sea push through slim so thick it's like an icebreaker going through the ice fields in Antarctica.

O.S. VOICE-OVER:

After decades of abuse, the oceans are in the last stages of a painful death, overcome by toxic waste and chemical spills. The warming of the oceans has led to smaller fish coming closer to shore. Those were followed by the predators who feasted on bathers as well. Partial bodies litter the sandy beaches: gruesome sights. The waste has created an atmosphere filled with toxins. Many fishermen, dock workers, sailors, and tug boats crews have died, or have been severely sickened before they realized the dangers.

In the final throws, the oxygen the oceans made - which was 50% of all the oxygen on the planet - was depleted leading the world to the brink of the greatest mass extinction in human history...

INT. CNBC STUDIOS. ENGLEWOOD, N.J. DAY

HOST KURT DEVLIN, silver-haired, polished demeanor, smooth talker, sits with his guest: **DR. ALBERT TASKER,** somewhat disheveled, out of place in a TV studio rather than his lab. Kurt makes the introductions to the TV audience, then turns to his guest.

KURT DEVLIN

Can you tell us simply doctor, what is happening to our oceans.

DR. TASKER
(belligerent, authoritarian)

No.

KURT DEVLIN

No? As in you can't tell us at all, or you can't tell us simply.

DR. TASKER
(waving his hand dismissively)

It took me forty years to understand how the oceans and the climate and photosynthesis interact. And you expect me to explain it in a few sentences to your audience comprised of lobotomized insects who can barely read or write?

KURT DEVLIN

That's a little cynical doctor. Didn't Einstein say that if a scientist couldn't explain a concept to the layman that meant he didn't understand it himself?

DR. TALBOT
(stunned, on the defensive)

Well...I...er, yes, that's true. Very well. Most living species need oxygen to keep their cells alive. Sea creatures survive on the oxygen that is by-product of photo-synthesis. If there are a large number of plants and fauna in marine waters, then oxygen levels can be quite high. Fewer plants, or fewer plants receiving sunlight, mean less oxygen. Oxygen is measured in parts per million. Levels can range from zero to over twenty.

3

                    KURT DEVLIN
                        (leans in)
            And now...?

                    DR. TALBOT
                    (shrugs his shoulders)
            Zero. Or so low as to be un-
            measurable. The waste is covering
            the marine plants so they can't
            receive sunlight and all the fish
            and mammals are dying.

                    KURT DEVLIN
            And so what can we do?

                    DR. TALBOT
            We can watch the oceans die. And
            when that happens, we, as a species,
            will be next.

                    KURT DEVLIN
                    (cannot fathom the
                     doctor's conclusions)
            Surely there's something...

                    DR. TALBOT
                    (shakes his head,
                     "The fools")
            At this point, it would be easier
            to build a space ship to take us
            to another planet, where we could
            ruin another pristine world, than
            it will be to save the oceans on
            Earth.

EXT. THE U.N. NEW YORK CITY. DAY.

A PANORAMA SHOT from a helicopter descending as it
circles the magnificent structures of New York City:
THE WORLD TRADE TOWER, THE EMPIRE STATE BUILDING and
**THE BROOKLYN BRIDGE, ZOOMING IN ON THE UNITED
NATIONS BUILDING.**

INT. THE U.N. GENERAL ASSEMBLY. NEW YORK CITY. DAY.

The great hall is packed; members huddle together on the floor level while guests edge forward in their mezzanine seats. The air is electric, all the representatives from the 193 foreign countries know what is at stake.

The Secretary General (**SIR NIGEL CAMERON**) of the organization addresses the audience comprised of the leaders of every country in the world. Nigel is aristocracy, regal, elegant, mannered, airy.

The mood is solemn. Only reserved conversations take place among some of the representatives.

Nigel Cameron taps his mic. He waits decorously for order to be restored.

                    NIGEL CAMERON
                 (losing patience)
          Ladies and gentlemen. May I have
          your attention, please. We are all
          here in a common purpose: to save
          the oceans and thereby save
          ourselves.

There is brief, reserved clapping from the attendees.

                    NIGEL CAMERON (con't)
          To the teams and proposals that did
          not make the final round of judging,
          we give our sincere gratitude.
          Other suggestions, that we
          institute policies and limitations
          on pollution, carbon emissions and
          oxygen-depleting chemicals are too
          little too late. We do not have
          decades or centuries left, ladies
          and gentlemen. We have weeks and
          months.

Knowing **NODS** and **MURMURS** between members.

> **NIGEL CAMERON (con't)**
> We are beyond debating, beyond
> personal agendas. The world, and we,
> are at a precipice. Our purpose
> here today is to decide which of
> the four most promising approaches
> we will choose and which one we
> will give our blessings and
> economic support to.

More nods in agreement, more intense, last minute
conversations.

> **NIGEL CAMERON (con't)**
> Will the competing teams please come
> to the podium to be recognized.

ANGLE ON: The teams, seated in the first row, step
on to the stage, waving and smiling for the
audience.

BACK TO SCENE: Nigel nods proudly at the entrants.
Taps his hands together lightly. As Nigel announces
the teams, they stand and wave to the audience.

> **NIGEL (O.S.)**
> From Russia, team leader, **SERGI
> LOVRIN** and **ANASTASIA LEVNAYA**.
> From Japan, team leader **SOSHI YAKU
> and TOMOE GOZEN**. From England,
> **RONALD DAVIES** and **PAULA MEYERS**.
> From America, **ADAM** and **EVE REARDON**.

BACK TO SCENE:

The audience claps resoundingly for the teams.

> **NIGEL CAMERON (con't)**
> You have, previously, heard all
> four presentations, and shared
> those with your scientific
> communities. It is time to vote on
> which team will lead the effort to
> save our oceans. You may now press
> the button by your seats, indicating
> one for Russia, two for Japan, three
> for Great Britain and four for America.

MURMURS increase in frequency and decibels as some member speak hurriedly with their subordinates, then nod their heads and press their buttons. The results tally on a **LARGE LED SCREEN** in back of the secretary general. Several nervous minutes go by before all the countries vote.

Clearly visible to all is that team Britain wins, with the U.S. second, Russia third and Japan fourth.

Sir Nigel puts a hand out and urges team leader Ronald Davies (medium height, pale, unkempt brown hair, arrogant, confident) **SASHAYS** to the podium.

> **RONALD DAVIES**
> Thank you all for placing your
> confidence in me. I...

Ronald turns and nods dismissively to his partner, Paula Meyers.

> **RONALD DAVIES (con't)**
> ...and Paula do not take lightly
> the task at hand and we pledge our-
> selves to validate your faith in us.

EVE REARDON (pixie cute, sexy even while trying to hide it, gutsy, outspoken, passionate to a fault about her work) is bursting at the seams. She begins to stand. Her husband, **ADAM REARDON (**edgy from growing up poor, street smart, boxer tough, ruggedly handsome. Would look more at home at a construction site) pulls Eve back into her seat.

> **RONALD DAVIES (con't)**
> There are those naysayers...
> (he nods at Eve and
> sniggers)
> ...who believe that our microbes
> will not do their job. To those
> I say, look at the results of the
> 2010 BP oil spill.

Ronald presses a remote and a large screen descends from the ceiling.

RONALD DAVIES (con't)

The spill in the Gulf of Mexico
would have been far worse if it
were not for microbes which did a
magnificent job of eating the oil
and then, ate dispersants as well.

Eve jumps to her feet. She takes a threatening step
toward Ronald who stutter-steps backwards.

EVE REARDON

Natural gas, Ronald; the microbes
ate the natural gas part of the
spill, not the oil. Microbes are
element specific. The human waste
is multi-faceted. Further--.

Nigel points a rude finger at Eve.

NIGEL CAMERON

Ms. Reardon, you will please sit
down and allow Mr. Davies to
complete his presentation.

Ronald nods conspiratorially at Nigel.

RONALD DAVIES

The only defense against the ongoing
Deepwater Horizon oil spill in the
Gulf of Mexico was billions of
hydrocarbon-chewing microbes.

ADAM REARDON
(stands, vents)
The microbes could only eat the oil
after it was broken down by the
dispersants; millions of gallons of
it. You're not using any dispersants.

RONALD DAVIS

We, Paula and I, have modified the
microbes to eat larger molecules.

Adam breaks out laughing. Eve places a hand over his
mouth and turns his head towards her. He gets the
message.

8

ADAM REARDON
(jabbing a finger at
Ronald)
You haven't even taken samples of
the waste from all five oceans.

RONALD DAVIS
(dismissively)
We extrapolated.

ADAM REARDON
The waste might cause the microbes
to mutate.

RONALD DAVIS
(waving his hands)
Yes, yes. We know that.

ADAM REARDON
Do you know, or care, how the
microbes might alter?

RONALD DAVIS
Well, that is...I

ADAM REARDON
The waste may have mutated, grown
so large the microbes can't consume
it. Or it may turn the microbes
themselves into waste, waste that
no microbes can attack.

RONALD DAVIS
Your jealousy is distorting your
logic, Adam.

ADAM REARDON
(demeaningly)
What about the viruses, Ronald?

RONALD DAVIES
Viruses?

ADAM REARDON
Yes, the ones that can infect the
spill-eaters. Creating a virus
impervious to any known antibiotic.

RONALD DAVIES

Yes...well, we have tested our
microbes against a wide spectrum
of virus and the microbes did not
succumb to those.

Eve stands, takes a step in Ronald's direction.
Adam restrains her.

EVE REARDON

Yes! That's the point, Michael. No
one can test for all the possible
ways the microbes can and will evolve.

RONALD DAVIES

And so you suggest we sit back and
do nothing?

EVE REARDON

I'm only saying we need to try the
safer methods before we commit our-
selves to an irreversible path.

RONALD DAVIES
(laughs haughtily)
In case you weren't aware, we will
quickly pass the point of no return
without a totally new approach.

EVE REARDON

At least tested, Michael. You
know, computer simulations, small
scale, real-world experiments.

RONALD DAVIES
(indignant)
We have run simulations.

Now Adam gets up, takes a threatening step forward.
Eve restrains him.

ADAM REARDON

I saw you work. You didn't allow
your simulations to run out. You
stopped where they successfully
consumed the waste but didn't
let them run their course.

Ronald carefully noted the distance between himself and Adam.

> **RONALD DAVIES**
> The experiment was over when the waste was gone. The microbes were inert; they dissipated afterwards.

Adam take a step toward Ronald, a hand balled into a fist. Eve reaches up, pulls him back into his seat.

> **ADAM REARDON**
> You don't know that, you dunce The microbes will interact with not just inert substances, living creatures will feed on it. Either the microbes or the feeders could evolve.

> **EVE REARDON**
> ...or devolve.

> **NIGEL CAMERON**
>> (hands out in front of him, beseeching for peace)
> Please, my friends, let's not revert to personal attacks.

> **EVE REARDON**
> You're insane, Ronald. Your meth-odolgy will unleash a plague that will dwarf the problems will already have. So far, microbes have been used to chew up only inert substances comprised of simplistic atomic structures.

> **RONALD DAVIES**
> The only thing out of control is you, Eve.

Adam gets up, face puffed up with anger, ready to box.

ADAM REARDON

That waste has been interacting
with fish and other waste. We
don't know what the molecular size
the waste is now and we don't know
if the combinations have produced
a whole new species.

RONALD DAVIES

It's naysayers like you that have
set back scientific discovery a
hundred years. The same warnings
were used against Copernicus, Galileo,
Linus Pauling, even Einstein.

EVE REARDON

You're too modest, Ronald, placing
yourself in such company only lowers
your stature.

RONALD DAVIES

Well, thank you for--.

ADAM REARDON
(jabbing a finger at
Ronald)
If you're so sure of your test
results, how about you take a swim
with some waste and your microbes
and see what happens.

RONALD DAVIES
(fighting back)
I'll have you know--.

ADAM REARDON

You'll have me know nothing.

Ronald Davis, chest puffed out, steps into Adam's
space.

Adam wraps a hand around Ronald's arm and squeezes.
Ronald bends down from the pain. Adam leans in and
whispers.

ADAM REARDON (con't)
Listen you nit-wit. It was only
though Sir Nigel, who happens to
be a relative on your wife's side
that you won the funding.

NIGEL CAMERON
(stands up, comes
between the two men)
That will be quite enough, gentlemen. We will maintain a proper
level of decorum.
(turns to Ronald Davies)
As per the parameters of the
program, you will keep this council
informed as to your progress and
aware of any set-backs or
complications.

EVE REARDON
(livid, seething)
When, not if.

WHITE LETTERS. BLACK SCREEN: ONE WEEK LATER

EXT. PORT AUTHORITY, RED HOOK TERMINAL. BROOKLYN, NY

A grand, ribbon cutting ceremony with the mayor of
NYC: **ZACHARY O'DONNELL** (corpulent, bald, bursting
out of his size 55 suit) and the Governor: **SAMUEL
TUSK** (a scarecrow of a man). Also officials of the
port and other dignitaries and a shit load of press.

ZACHARY O'DONNELL
(beaming for the press)
We are gathered here today at these
docks at a critical time for our
country and our world.

Grandiose applause greets the mayor's intonation.

ZACHARY O'DONNELL (con't)
Thank you all for that warm welcome,
but it is not I who should be the
center of attention. Rather, it is
the man and woman to my right:
Ronald Davies And Paula Meyers.

Michael steps forward, effectively cutting off
Paula, who is exasperated but used to the behavior.

Ronald grabs the mic from a startled mayor and
addresses the gathering.

> **RONALD DAVIES**
> Thank you, Mr. Mayor. Today is a
> momentous day for all of humanity.
> For today we begin the process of
> cleaning our oceans and preserving
> our environment.

Spontaneous applause breaks out. Ronald basks in
the glow, doing nothing to quell the crowd. Finally
he waves his hands to restore order.

> **RONALD DAVIES (con't)**
> We, my partner, Paula...
> > (he mentions in passing)
> ...have created an organism that
> will consume human waste and turn
> it into an inert substance. There
> are those skeptics who preach
> danger and havoc, but the precedent
> of using microbes to clean environ-
> mental disasters is were documented,
> specifically Deep Water Horizon in
> the Gulf of Mexico.

> **REPORTER #1**
> > (stands up)
> What about the possibly of microbes
> using the waste to mutate?

> **RONALD DAVIES**
> Did the microbes in the gulf mutate
> when they were exposed to the
> hydrocarbons?

> **REPORTER #2**
> Every spill: whether it be oil or
> sulfur from paper mills, or tail-
> ings from iron ore mines, or
> methane from fertilizer factories
> has required a different dispersant.

How can--.

RONALD DAVIES

May I ask what institution of
higher learning you attained your
degree in microbiology from?

REPORTER #2

It doesn't take a rocket scientist
to realize the dangers involved
here.

Zachary nods to security who start to make
their way over to the reporter.

RONALD DAVIES

I am a rocket scientist, you moron.

ZACHARY O'DONNELL
(hands out in front
of him)
Now, now, gentlemen; there's no need
for name calling. Let us proceed
with a modicum of decorum.

RONALD DAVIES
(to reporter #2)
And so you would suggest...?

REPORTER #2

More testing.

The security team reaches the reporter who
tries to hold his place.

RONALD DAVIES

While the oceans die and it becomes
too late for any process to reverse
the situation?

REPORTER #2
(struggles against
the men)
That's not--.

The security personnel finally drag the
reporter away.

Ronald Davies nods his head at the mayor. The Mayor nods at a longshoreman who pressed a button which lowers the cage containing the microbes into the water.

The gate opens and the microbes float out into the ocean.

MONTAGE:

The scene is repeated in New Orleans, Vladivoctok, Jakarta, London, San Paulo, Manila, Johannesburg, Shanghai. In all countries, crowds are cheering, clapping, waving placards: **"BRING BACK OUR SEAS."**

BLACK SCREEN.   WHITE LETTERS: 30 DAYS LATER

EXT. NOVA SCOTIA. 30 MI. EAST OF HALIFAX. DAY

A small, two-man fishing boat.

Fishermen cast their nets into the darkened sea. The captain of the vessel motors slowly ahead, dragging the nets behind the boat, scooping up their catch.

After a a short time, fisherman #1 (**FRANK**) uses a hand crank to raise the nets.

Frank is using all his strength to turn the crank.

> **FISHERMAN #2 (TONY)**
> (laughing his ass off)
> Whatsa matter, Frank. Didn't get
> your coffee this morning?

> **FRANK**
> It ain't me, dickwad. We musta
> landed a Great White Shark.

Tony walks over carefully to Frank.

> **TONY**
> Lemme help you before you start
> cryin'.

The two men double up on the crank. Slowly the catch is brought on board. They use hooks to sift through the waste and get to the catch. Something slithers up out of the slime and attacks Frank. He cannot move, frozen in place either by fear or something very powerful. His body begins to shake. He drops his hook. His body is lifted into the air by something from below.

And then the head of a serpent comes out of Frank's mouth, Frank's **HEART AND KIDNEYS AND LIVER** dangling from its jaws.

The head is part monkey, part human, part snake. Gruesome enough to kill a man just by looking at it.

Its mouth is that of a carnivore, two sets of long, sharp teeth.

Just as silently as the creature entered Frank's body, it slinks back down, allowing Frank's lifeless, empty body to collapse to the ship's deck like a deflated balloon.

Tony, startled at first, unsure if what happened really did happen, crosses himself, then walks over to the side of the boat where the sea monster retreated into the depths.

As Tony looks into the dark, bracken ocean, a tail subtly wraps around his leg, so softly, there is barely any pressure. And then the creature tightens the noose and drags Tony down into the dark sea.

INT. CNBC. ENGLEWOOD CLIFFS, NJ. EVENING

Moderator **GARY ARNOLD**: suave, polished, confident, is flanked by two guests: Ronald Davies and **PROFESSOR EMERITUS, MARTIN LOVEJOY**: rumpled, a big shock of frazzled white hair, haphazardly dressed, bifocals on the tip of his nose.

> **GARY ARNOLD**
> Welcome Doctor Davies and welcome
> Professor Lovejoy. We are honored
> to have you on our program to
> discuss the merits of the recent
> actions taken to save our oceans.

17

RONALD DAVIES
Thank you, Gary. I would like to say how grateful I am for the opportunity to--.

PROFESSOR LOVEJOY
(waves a dismissive
hand)
Your plan hasn't worked yet you twit. Nothing has happened and there is just as good a chance that something horrible will occur as there is that the ocean will be saved by your concocted soup of microorganisms available at Wal-Mart.

RONALD DAVIES
(leaning forward)
I'd like to respond to that.

PROFESSOR LOVEJOY
You have already spoken. You think your little friends are out there in the big oceans gobbling up waste.

RONALD DAVIES
If I may, professor--.

PROFESSOR LOVEJOY
No, you may not, sir. You had your chance to speak numerous times between when your little Frankenstein monsters were released into the seas and now.

RONALD DAVIES
I was hospitalized with a bug I caught while doing research in South America.

PROFESSOR LOVEJOY
Well, I'm afraid it's too late now, doctor.

RONALD DAVIES

That is one man's opinion.

PROFESSOR LOVEJOY

It's a function of place and size, and we don't know how much waste is out there. Without that information you can't begin to make any kind of calculation of how may genetically modified microorganism would be needed. And how did you expect to keep the microorganism together with the waste?

RONALD DAVIES

Using aerial photos and satellite images we have--.

MARTIN LOVEJOY

You nincompoop. You don't know shit from Shinola.

Gary Arnold gets partially out of his seat and reaches out a hand to restore order.

GARY ARNOLD

Now, now, gentlemen, there's no need for--.

PROFESSOR LOVEJOY
(to Ronald)
You'd be better off using dish detergent, you moron.

RONALD DAVIES

We have gone to great lengths to—

PROFESSOR LOVEJOY

Tell me, is one of your agents Corexit 9527A?

RONALD DAVIS

Yes, but how...?

PROFESSOR LOVEJOY

Corexit 9527A contains the solvent 2-butoxyethanol, which is a known human carcinogen and toxic to animals and other life.

RONALD DAVIES
(dismissively)
Yes, we are aware of that.

PROFESSOR LOVEJOY

Then are you aware that the waste will linger for a long time. The microbes break down hydrocarbons in weeks to months to years, not hours or days. Much of the tar or asphalt compounds are not readily subject to microbial attack. Tar tends to persist, as does asphalt.

RONALD DAVIES

You are not saying anything we haven't already considered.

PROFESSOR LOVEJOY

Well, sir, did you know that the waste now in the ocean is comprised of thousands, maybe millions of elements, and compounds made from those elements are creating substances we have never seen before...and have no way of knowing whether your microorganism will kill them or help them grow.

RONALD DAVIES
(dismissively)
Time was of the essence. We did not have a chance to--.

PROFESSOR LOVEJOY

And what about the lack of oxygen, required for all chemical reac-tions. You must know thae there are only one or two parts per million oxygen in the oceans.

20

RONALD DAVIES
(flustered)
Well, er, yes, that's true, but when the microorganisms begin eating the waste, photosynthesis will start. And at that point--.

PROFESSOR LOVEJOY
If, doctor, if the microorganisms can eat the waste without oxygen, something science says is impossible.

EXT. / INT. HOME / LAB OF ADAM AND EVE REARDON. DAY

A modern, predominately glass home on the edge of a lake with ducks, geese and fish. An idyllic setting.

The two are surrounded by beakers of various substances, electron microscopes, cyclotrons, walk-in freezers and 4 large screen computers.

EVE REARDON
(stressed but hopeful)
What do you make of the arguments against Ronald's army of micro-organisms, Adam? They can't all be wrong.

ADAM REARDON
(assuredly)
When certain forces are set in motion, the action will never stop exactly where the planners hope or project. We're talking about eco-systems. It's impossible to conceive of a situation where there's such a massive change to the seas and there's not either a reversal or a cancerous, or an on-going mutative action.

EVE REARDON
(stumped)
Do you have any idea what that mutation might look like?

ADAM REARDON

I was going to ask you. You're
the theoretical bio-physicist.

EVE REARDON

I've done simulations, but I'm
having a tough time believing what
I'm watching.

ADAM REARDON
(anxious, pressing)
Let's see them.

EVE HANSEN

My projections look like is something out of a science fiction
movie.

Adam leans over to her, staring at the large
computer screen.

EVE REARDON

I started with the microbes David
began with. I then added the chemical
make-up of the wastes we know of. The
microbes didn't mutate, rather they
simply dissolved after consuming
the substance they were bred to go
after. But...?

ADAM REARDON

But...

EVE REARDON

Then I combined the elements in
one form of waste with elements
in another form of waste and
allowed those elements to stew for
a number of generations or so.

ADAM REARDON

And then?

Adam Reardon leans over Eve's shoulder, trying to
focus on the computer screen but distracted by the
scent of her.

Eve turns around, sees what Adam is up to. She
smiles, then laughs, before getting back to
business.

> **ADAM REARDON**
> (caught in the act)
> Sorry. Show me.

Eve runs the computer simulation.

ANGLE ON: EVE'S COMPUTER SCREEN.

The microbes start to grow, slowly at first.

> **EVE REARDON (voice)**
> The 'X' axis is size, the 'Y' axis
> is time. By six weeks, the size
> goes parabolic.

> **ADAM REARDON (voice)**
> But that's still relative to the
> size of the original microbe.

The grow chart go **PARABOLIC.**

> **EVE REARDON (voice)**
> Twenty-six feet, Adam.

BACK TO SCENE:

> **ADAM REARDON**
> Jesus, Eve.

> **EVE REARDON**
> I don' think God plays any part
> in this. And it may even be
> beyond the Devil's purview.

BLACK SCREEN: THREE MONTHS LATER

EXT./INT. WASHINGTON D.C. WHITE HOUSE SOUTH LAWN

A proper ceremony masks a celebratory undertone.
President **EDWARD OWENS** pins the **MEDAL OF FREEDOM** on
RONALD DAVIES and PAULA MEYERS.

After pinning the medals, President Owens shakes hands with both scientists. The reserved crowd of invited guests breaks out in spontaneous applause. An aurora of victory permeates the gathering.

INT. CNN NEWS DESK, ATLANTA, GA. EVENING.

One of the biggest, busiest news desks in the world in prime time. **TRENT NELSON:** young, aggressive, smart, a man coming up fast.

> **TRENT NELSON**
> What seemed like isolated incidents have now been pieced together to give a very disturbing picture. Excitement over observations of the seas turning blue have coincided with creatures called giant eels by some, snakes by others, creature from hell by still others. What at first were attacks on smaller targets: fishing vessels, swimmers, divers have escalated to yachts and freighters, cruise ships and ocean-going vessels, Many vessels has simply gone missing, others, on GPS and auto-pilot have reached shore with no crew or passengers on board. People have been warned to stay out of the ocean and away from the shores. But international trade must still go on. The largest vessels have been fitted with guards with machine guns.

MONTAGE:

NEWS ANCHORS in Brazil, Panama, Spain, The Philippines, Indonesia, Italy, Greece, South Africa tell the same story while scenes of the actual happenings unfold.

GIANT LIZARDS are slithering on to boats. People are being dragged back down into the depths.

BATHERS: children, men, women are sucked under the waves, gone so fast others aren't sure they were there in the first place.

24

BODIES wash up on shore, minus their internal organs.

KAYAKERS on the oceans paddle their way out to sea. Until...a huge snake-like creature wraps its twenty foot tail around the kayak and drags it down before the people can react.

DOCK WORKERS loading and unloading cargo just feet from the water. Something grips their legs, pulling them off the docks and under the dark water before they could even scream.

PRIVATE YACHATS are dragged under the water.

CRUISE LINERS are attacked. When night approaches, the lizards use the ropes - which hold the life boats, and ladders used by passengers needing to abandon the ships - to climb aboard.

The huge, ocean-going vessels arrive with no crew or passengers on board.

EXT. PERTH, AUSTRALIA. DAY

A surfing competition on the Great Barrier Reef. Dozens of surfers lined up, ready to catch a big wave. Just as it crests, the creatures appear and snatch the surfers off their boards. The boards continue on, catching the wave and riding in to the shore. All the audience sits stunned, thinking what they saw was their imagination and that the surfers just lost their boards. The lizards are snaking through the dead bodies, down the neck, out the back side, eating the vital organs which drip from their mouths.

MONTAGE:

Scenes from every major city bordering the oceans - Guangzhou, China. New Orleans, America. Guayaquil, Ecuador. Ho Chi Minh City, Vietnam. Abidjan, Cote d'Ivoire. Zhanjing, China. Mumbai, India. Khulna, Bangladesh. Palembang, Indonesia, and Shenzen, China - show people fleeing for their lives.

People flee in cars in the rich nations, others, from poorer nations ride on mules, horse carts, or go on foot. Scenes of car accidents; scenes of people being trampled, scenes of looting, fighting in the streets, killings.

EXT. THE COAST LINE OF MAJOR CITIES IN THE WORLD.

Silently, swiftly, the lizards swim through the effulgence being pumped into the oceans, making their way into the sewer systems of those cities. The number of creatures migrating is astronomical. Wave after wave of the things slip into the pipes.

BLACK SCREEN, WHITE LETTERS: SIX MONTHS LATER. DAY.

INT. BBC. International News Desk.   London.

A modern, sleek, glass and chrome newsroom.

ANCHOR **JASON SAUNDERS**, smooth, implacable, unflappable, shuffles his papers in order. Trent looks to the camera man who is counting down from ten. The studio lights change from red to green. Trent nods to the cameraman and begins his broadcast.

> **JASON SAUNDERS**
> Welcome viewers. Reports are coming
> in from all over the world
> indicating that the seas are
> changing. There have been numerous
> sighting of blue patches opening up
> in the Atlantic, Pacific, Indian,
> Southern and Artic Oceans —all
> reporting the same phenomenon.

VISUALS intersperse with Jason's report:

ANGLE ON: Scenes from all five oceans show bright patches of blue in the otherwise black oceans.

The scene continues as the reporter continues.

> **JASON SAUNDERS (O.S.)**
> Coastal cities are returning to
> normal after being almost emptied
> due to the threat of so-called sea
> monsters. The scientific community has
> proffered alternative explanations,
> sighting mass hysteria, schools of
> whales, great white sharks in
> greater numbers and closer to shore
> due to the warming of the oceans.

ANGLE ON: FISHERMEN'S CATCHES growing. SURFERS AND SWIMMERS returning to the waters. FISH are visible even six feet below the surface. CORAL REEFS beginning to grow. WATER SKIERS out in force. PARENTS, CHILDREN IN TOW, venture out to the first set of breaking waves. People on shore rejoice. SEA BOAT CAPTAINS honk their air horns.

INT. THE HOME OF DRS. EVE AND ADAM REARDON. EVENING.

Adam is watching the news of the oceans returning to health.

Eve is in the bedroom seated at a make-up mirror. She is bald, brushing two wigs that sit on Styrofoam heads. She is debating which one to put on.

Adam calls from the living room.

> **ADAM REARDON (O.S.)**
> Honey, you should come and watch
> this.

With a **TEAR** in her eye, Eve chooses the short, blond wig and fits it in place. She puts on a happy face before entering the living room.

She stands behind Adam who remains seated, her hand on his shoulder. She is crying and doesn't want Adam to see, but he can sense it. Adam draws Eve around from in back of the sofa and pats the cushion next to his. She sits and squeezes herself as close as possible to Adam.

                              ADAM
              I love that hair style on you, dear.

                               EVE
              Even if it's not mine?

                              ADAM
              Yours will grow back, sweetie.

                               EVE
              No, it won't, Adam. You need to
              face up to this. I'm going to die.
              You need to get on with you life.
              I can help you meet a good woman.

Adam gruffly pushes Eve away so he can look here in
the face.

                              ADAM
              I don't want to hear that kind of talk.
              Staying positive is the most important
              thing sick people can do.

                               EVE
              I'm not sick, Adam. I have cancer
              and it's a death sentence.

                              ADAM
              I'll kill you myself if you keep
              saying that.

EXT. ROW HOME IN RUST BELT CITY. NIGHT

The home is a simple one: living room, dining room,
kitchen on the first floor; bedrooms, bathroom on
the second floor.

TWO PEOPLE making out like crazy on the sofa. The
WOMAN (voluptuous, a killer body, breast and ass to
die for) is down to her panties. The MAN (hard
body, some tattoos, six pack) is shirtless with his
trousers open but underpants still on. The man is
turned on; almost out of control.

                                28

                            MAN
            Well, are you finally out of
            excuses?

                           WOMAN
                    (playful)
            I know, I know, butt sex, you
            don't have to try to convince me.

She gets up, revealing more of her body. We have a
side view showing off her awesome, full breast and a
killer ass while not revealing her bush.

                            MAN
            Then where are you going?

                           WOMAN
             I'm going to 'freshen up.'

The man is playing with himself through his
underwear.

                            MAN
            What does that mean?

The woman walks upstairs to the bathroom, her ass
jiggling like a taught drum. Mesmerizing. The woman
turns back, smiling conspiratorially.

                           WOMAN
            You know...flush.

                            MAN
            Oh, yeah.

The woman walks down the hall to the bathroom,
closes the door and sits down on the toilet.

She picks up a **'PLAYMATE' MAGAZINE** and fantasizes.

                         MAN (O.S.)
                    (losing patience)
            Hey, aren't you done yet!

                           WOMAN
                    (she giggles)
            Don't rush me.

                             29

                    MAN (O.S.)
          I'm gonna start without you.

                     WOMAN
          Hold your horses, mister.

The woman is every bit as turned on as the man. But
then there is a change in her demeanor. At first
she is surprised, then shocked, then afraid.

She tries to get up but is unable. She grabs for the
towel rack and pulls but the rack breaks off.

Her body begins to shake violently. She is slightly
lifted off the toilet. She **SCREAMS**.

                    MAN (O.S.)
                 (laughs, yells)
          Hey, we haven't even started yet.

Her mouth opens impossibly wide. And then the
snake-like creature comes out of her mouth, her guts
dangling from the thing's jaws.

Just as swiftly as the creature appears, it slithers
back down the drain. The woman's lifeless, empty
body collapse like an accordion.

EXT. / INT. LAB OF ADAM AND EVE REARDON.

The two of them are working intensely on their
individual computers, then compare their findings on
a third (larger) computer.

                  ADAM REARDON
                 (incredulous)
          So the sea serpents are gone and
          we're left with a blue ocean?

                   EVE REARDON
                 (unconvinced)
          Sounds too good to be true.

                  ADAM REARDON
          And no one is even questioning that.

                    EVE REARDON
              (smacks the desk
              with her hand)
         I can't replicate what David did.

                    ADAM REARDON
              (patting Eve on the
              shoulder)
         I'll bet you David couldn't
         duplicate what he did. If he could,
         we could reverse engineer the
         microbes and experiment on them.

                    EVE REARDON
         The people have been lulled into a
         false sense of security. Creatures
         who reeked such havoc simply don't
         disappear and are never heard of
         again. It's way too good to be true.

                    ADAM REARDON
         Everybody in the world but us is
         celebrating.

                    EVE REARDON
         We don't know how they replicate or
         how fast they birth.

                    ADAM REARDON
         They might be evolving. Each
         generation, even if only a few
         weeks old, may already be more
         intelligent, more impervious to
         biogenic engineering.

INT. LARSON COLLEGE. BETA ALPHA FRATERNITY. NIGHT.

Kegs of beer sit in every corner. Boys and girls,
half naked, run from room to room, screaming,
laughing, stripping.

There are waiting lines for the "Johns." People are
squeezing in their intestines, trying to stem the
flow.

A gorgeous blonde, (**COED #1**) Victoria's Secret underwear, the face and body of a goddess, is four back in the queue.

When the occupant of the bathroom comes out, the girl brushes past the others, jumps into the bathroom and locks the door. Immediately, the others, pissed, start banging in the door, cursing the girl.

She laughs so hard, she pees her pants. She throws them away and squats down on the toilet.

As she pees, a sense of relief spreads over her face.

But then her body jerks. Her eyes open wide, she tries to get up but is held down. She screams a blood curdling **SCREAM**.

At first, the young men laugh, thinking is frat house humor. The **SCREAMS** continue.

A giant creature slithers up her body from the basin and out her throat, her intestines dangling from its jaws.

The screaming stops.

The men grow afraid and push on the door. The door starts to give but is held by the chain. One of the men (**BOY #1**) tries to squeeze through the opening. Something grabs his leg and pulls but he is far to bulky to go through small opening.

The boy **SCREAMS**, his face puffed out with fear, turning purple in color.

                    BOY #1
                 (crying)
             Help me!

His body starts to **SHAKE VIOLENTLY**. The other boys **PULL** on his body. And then the head of the serpent comes out of the boy's mouth, baring its teeth and hissing like a snake.

The other boys and girls jump back and fall on their asses as they watch the snake withdraw from the body, taking with it the boy's internal organs, leaving just the skin of the boy in place. And then the rest of the body, now deflated, is easily dragged though the opening.

INT. CCTV. CHINA NEWS. BEIJING. EVENING.

News Anchor **TOMMY CHIN** has on a grim expression. He is very young, representing his newly emerged country. Very clean cut. Impatient with others who think slower than he does.

The **CAMERAMAN** signals a "GO."

> **TOMMY CHIN**
> Tonight we find ourselves at war
> with an enemy who strikes swiftly
> then disappears just as quickly.
> No one has learned where the beasts,
> once dismissed as aberrations come
> from. Scientists and politicians
> who told us the creatures were
> figments of our imagination are now
> back-tracking to cover their
> ignorance   and arrogance. And now
> the creatures have resurfaced on
> our shores. It has even been rumored
> that there have been attacks on homes,
> pedestrians and businesses. We must
> now ask ourselves if these attacks
> are coordinated or if they are even
> from our world. Scientific and
> military experts have gathered
> to determine how to fight this
> entity. So far, their efforts have
> gone for naught.

Tommy Chin turns to his guest.

> **TOMMY CHIN (con't)**
> We have with us tonight, Dr. Charles
> Yang, an expert on prehistoric
> species.
> > (smiles warmly)
> Welcome Dr. Wong.

The doctor nods demurely.

                    TOMMY CHIN (con't)
          What do you make of all the reports
          coming in from Chinese coastal
          cities and other cities around the
          world, of prehistoric monsters
          roaming the land.

                    DR. YANG
          Group insanity.

                    TOMMY CHIN
          But surely you can't dismiss so
          many reports outright?

                    DR. YANG
          The creatures described died
          off sixty-two million years ago,
          ending the Pleistocene Era when a
          meteor crashed into the Gulf of
          Mexico.

                    TOMMY CHIN
          But there is film, doctor. We have--.

Doctor Yang waves a dismissive hand.

                    DR. YANG
          I've seen better special effects in
          the last Godzilla movie. Do you
          think that monster is real because
          you saw it on film?

                    TOMMY CHIN
          Is it not possible that the
          creatures we are seeing today could
          be a newly evolved species?

                    DR. WANG
          Ha! It would take millions of years
          for such a reptile to evolve from
          those currently inhabiting the Earth.

                    TOMMY CHIN
          But maybe some unknown force of
          chemistry is at work here?

                    DR. WONG
          The only thing at work here is mass
          hysteria.

                    TOMMY CHIN
          Well, professor, let's have look
          at what some of our reporters filmed
          just hours ago, shall we?

ANGLE ON THE BIG SCREEN:

A group of lizards come out of the sewers and chase
pedestrians down the street.

BACK TO SCENE:

                    TOMMY CHIN
          What do you think now, doctor?

                    DR. WONG
          Bah! Hollywood theatrics. My film
          class at the university can do better.

                    TOMMY CHIN
          Well, thank you, doctor. I am sure
          we will all sleep better knowing we
          are safe.

                    DR. WONG
          Did I ever tell you of the time I
          rappelled into a cave that went
          almost to the center of the earth?

                    TOMMY CHIN
          That's all we have time for tonight,
          doctor.

                    DR. WONG
          But I haven't--.

Tommy Chin pulls a finger across his throat and the
scene at the studio goes black

EXT. MIAMI BEACH. NIGHT.

The coast has been turned into a war zone with the military and police establishing themselves as the last line of defense against the onslaught expected at any moment. It looks like civilization's last stand. There is praise from the city managers but the people and the military remain on edge.

Military vehicles are lined up at the shore line: Humvees, jeeps, half-tracks.

Soldiers stand between and behind the vehicles. Some with flame-throwers, others with machine guns, grenade guns.

The line of soldiers and vehicles is focused on the area where the last attack took place.

Behind the military are stands erected by the press with large Klieg lights illuminating the night.

There are even news helicopters overhead; their lights illuminate the sea just beyond the breakers.

The scene is duplicated in L.A. Louisiana, Seattle, and San Francisco.

REPORTER #1 is broadcasting from the stands set up behind the wall of solders.

> **REPORTER #1**
> Tonight, the nation is prepared to end the threat to our peace and security...and even our very existence.

The solders tense up as the surf **ROARS**, a threatening sound.

> **REPORTER #1 (con't)**
> There is confusion running through the chain of command here tonight. While all eyes are set on the ocean, we are feeling a rumbling under our feet.

The ground begins to shake the news reporters' scaffolding. The reporters hold on to metal poles for dear life.

Officers and enlisted men alike are confused. Some look out at the shore, others look down at their feet.

The ground shakes hard enough to jostle the leaves out of trees and nudge vehicles a few inches.

ANGLE ON: Several reptiles slither out of the sewer system. They are immediately cut down.

CHEERS go up from the on-lookers.

The soldiers **HIGH FIVE**.

But theN more reptiles come out of the sewers. There is a fire-fight. Dozens of the monsters are killed.

The wave of attacks quickly ends but the soldiers stand at the ready; nervous, looking to their commanding officers for orders.

Everyone: news media, on-lookers, and soldiers are apprehensive, not knowing if the second wave of monsters was the last or just a prelude to more.

Time passes. Nerves are frayed but soldiers start to relax. Reporters interview officers on their strategy of success.

And then, just before dawn, the real attack begins. Hundred, then thousands of the creatures pour from the sewers. They have evolved further. Now they have legs, small, but strong enough to **RUN** rather than slither. Their faces are more human: like a baby crying. They coil and spring in the air.

The lizards attack the soldiers, forcing their way down their throats and out their backsides.

Others lizards **JUMP** into the trucks, humvees and jeeps. **BLOOD** comes spurting out of the vehicles.

On-lookers and the news media run for their lives.
But the creatures chase after them and catch up.

Lizards are forcing themselves down the mouths of
people and coming out their backsides dragging the
innards of the people along.

Pandemonium sets in as on-lookers run for their
lives. The officers order a retreat but their
soldiers have already fled.

A lizard **BITES THE HEAD OFF** a **GENERAL** spits it out,
then slithers down the officer's neck.

INT. LABORATORY OF ADAM AND EVE HANSON. NIGHT.

Both Adam and Eve are deep into their work: INTENSE.
They exude passion and concern and a twinge of fear.

Eve leans back from her table, rubs her temples.

                    EVE REARDON
So, the sea creatures evolved at the
intersection of Ronald Davis' microorganisms
and the toxic waste in the oceans

                         ADAM REARDON
                    (remains focused on
                     his work)
                We already know that much.

                         EVE REARDON
                They thrive in an environment where
                the global mean concentration of CO_2
                in the atmosphere is the highest in
                at least the past 800,000 years and
                likely the highest in the past 20
                million years. The increase has
                been caused by the burning of
                fossil fuels and deforestation.
                30-40% of the CO_2 released by
                humans into the atmosphere
                dissolves into oceans, rivers and
                lakes.

ADAM REARDON

So we created the perfect conditions
for these creatures to breed and
grow.

EVE REARDON

Exactly, but then, all of a sudden,
they disappear from the oceans and
make their way to terra firma.

ADAM REARDON

(looks up from his
microscope)
So why does a creature that is so
successful in one environment leave
that environment?

Eve's face lights up as they suddenly make progress.

EVE REARDON

They would if their environment
changes.

Adam gets up and joins Eve at her work station.

ADAM REARDON

What changed in the oceans?

EVE REARDON

Well, for one thing, they turned
from black to blue.

ADAM REARDON

That's not all that changed.

EVE REARDON

What then?

ADAM REARDON

The oxygen content.

EVE REARDON

Shit! That's right. How did we
miss that?

Eve turns away in disgust of her failings.

Adam moves over to her and touches her cheek,
turning her head to face him.

>   **ADAM REARDON**
>   Let's not be too hard on ourselves.
>   No ones ever dealt with these
>   situations before.

Eve is crying again; Adam tries to comfort her.

>   **ADAM REARDON**
>   (cuddling her)
>   Are you in pain, sweetie?

>   **EVE REARDON**
>   (wiping tears away)
>   It's not that.

>   ADAM **REARDON**
>   Then what?

>   **EVE REARDON**
>   I want to make a difference before
>   I go.

>   **ADAM REARDON**
>   You're not going anywhere.

>   **EVE REARDON**
>   (pushes away from him)
>   It's not helping me to pretend this
>   isn't happening. I've got to work
>   harder, faster.

Adam strokes her hair...gently. Eve is
distraught, beaten, on the verge of giving up,
giving in to the disease.

>   **EVE REARDON (con't)**
>   It wasn't supposed to be like this.

>   **ADAM REARDON**
>   (draws her in)
>   How was it supposed to be, sweetie?

                              EVE REARDON
               It was supposed to be forever.

                              ADAM REARDON
               But nothing lasts forever?

                              EVE REARDON
               I didn't mean us, I meant the work
               we leave behind: the cures, the
               prolonged health and safety of
               humanity.

Eve's tears flow on to Adam's shoulder.

INT. CNN NEWS DESK. ATLANTA GA. EVENING.

News Anchor, **BRAD JENNINGS,** a silver-haired,
normally unflappable, 50s gentleman views the
city scene at the same time as the television
audience does.

EXT. CITYSCAPE. RUSH HOUR. NIGHT.

An aerial view of downtown with its skyscrapers
surrounded by smaller edifices. A sea of
humanity rushing to and fro. Shoulders and elbows
touching, no room for personal expression.

A **SIREN** blares from every corner.

                              SIREN (voice)
               Residents within a half mile of the
               shore line must evacuate their homes
               and proceed to the Red Cross shelters
               set up inland. Visitors should stay
               away from the ocean. All citizens
               Should also give a wide berth to the
               storm drains.

EXT. ARIAL VIEW OF THE SHORE LINE.

Traffic is bumper to bumper for miles. Cars are
burdened inside and out, carrying as many
valuables as the vehicles can hold. Others,
without transportation, carry back packs, duffle
bags and suitcases. Reminiscent of Hurricane
Katrina and New Orleans.

                    BRAD JENNINGS
                         (astonished)
          My God! Is this happening in our
          country?

He puts a hand to his ear piece, listens, nods.

                    BRAD JENNINGS (con't)
          I am getting word that this scene
          is being repeated in many countries
          around the globe.

MONTAGE:

In coastal cities all over the world (Africa,
Southeast Asia, The Americas, Oceana) citizens
are retreating from the oceans.

BACK TO SCENE:

Brad Jennings takes a handkerchief and wipes his
face of perspiration and tears.

                    BRAD JENNINGS (con't)
          So far, no one, not scientists or
          the military have a strategy to
          combat whatever it is we face.

Brad taps his ear piece.

                    BRAD JENNINGS (con't)
          I've just been told we have a reporter
          at the scene of a recent attack. We'll
          switch now to Robin Meyers.

EXT. A CITY STREET SCENE. NIGHT.

A major urban setting: high-rise buildings,
kiosks on every corner, street lights glare in
the drizzling rain, throwing shadows, garishly
illuminating red taillights and blinding
headlights.

ANGLE ON: Many people are lying on the street, their blood coagulating next to them. Emergency vehicles are on the scene but can do little but cover the bodies. The coroner's van is on the scene, attendants loading bodies in the back. A priest gives last rights to some of the dead and dying.

A **CAMERAMAN** has Robin in his sights but is totally focus on her face and does not see any thing outside of that.

>   ROBIN MEYERS
> Brad, this is a disaster of epic pro-
> portions. All around me, not just dead
> bodies but their entrails scattered on
> the streets and on the sidewalks. I've
> tried to speak to the various police
> and health and safety officials but
> they don't know anymore than we do.
> We seem to be fighting an enemy that
> has the power to strike with deadly
> precision and just as quickly
> disappear.

From behind the reporter, a large, long shape moves out of the alley on its short legs. It is almost to the feet of the reporter when...

SIRENS WAIL in the background; the noise interfering with conversations and **ORDERS BARKED** by police and EMTs.

>   BRAD JENNINGS
>   (screaming, eyes bugging
>   out)
> Robin! Behind you!

>   ROBIN MEYERS
>   (tapping her earpiece)
> Brad? What was that? It's impossible
> to--

The serpent walks rapidly on its small legs until it is at Robin's feet.

It raps its tail around Robin's feet.

She looks down, sees the serpent, screams, tries to run but is held in place. The serpent enters Robin's body.

Robin screams until her voice gives out.

The serpent comes out robin's mouth, looking at Robin, smiling with its baby face, robin's intestine dripping from its jaws.

And then its jaws unhinge and its mouth opens impossibly wide, showing off two rows of jagged, sharp teeth.

Robin is shaking her head "**NO, NO.**"

The serpent vomits Robin's entrails which land at the feet of the cameraman who continues filming because that is his natural instinct.

The creatures slides down Robin's neck and comes out her backside. Robin's deflated body spills to the ground.

INT. TV STUDIO. NIGHT

Brad Jennings throws up into a trash can next to his desk.

EXT. CAMPGROUND. MIDNIGHT

A little used spot deep in the forest. No signs of urbanization or people as far as the eye can see and the ears hear.

A star-lit night, crickets abound in the background.

Six men and women are gathered around a campfire. Laughter, stories being told, each person trying harder than the one before to scare the others. Empty beer cans and liquor bottles attest to how much drinking has been going on.

CAMPER #1 (MAN, 30s)

Did you know how Frankenstein got written?

CAMPER #2 (WOMAN, 20s)

Yeah. Mary and Percy Shelly were visiting Lord Byron and they got stuck in a castle because a storm washed away a bridge and roads. They made up stories to pass the time.

CAMPER #3 (GIRL, 18)
(giggling)
What if a dentist was getting audited by an I.R.S. agent and the agent mentioned he had cavity. Then the dentist offers to check it out.

CAMPER #4 (GUY, 20)

That's not scary; that's stupid. Why would a guy help someone who's auditing them?

CAMPER #3

You didn't let me finish.

CAMPER #4

You can go ahead without me.
(farts loudly)
I need to take a dump.
CAMPER #2
(throws a stick at
Camper #4)
You're disgusting. You can take your sense of humor and your farts with you.

Camper #4 wanders into the woods, stumbling over branches, laughing like a hyena.

CAMPER #3 (O.S.)

So the dentist puts the I.R.S. guy in a chair and gives him a shot of Novocain. But it's really a knock out drug. When the government guy

> **CAMPER #3 (O.S.) (con't)**
> wakes up, his mouth is stretched so
> wide by clamps that his lips are
> bleeding. Then the dentists tells
> him that it's really serious and
> he'll have to drill out the cavities.
> So he starts drilling but doesn't
> give the guy Novocain and—

> **CAMPER #4**
> (runs back into the camp,
> screaming)
> Ah! Ah! It hurts!

All the other campers **SCREAM**, then recover and
throw empty beer cans at Camper #4 as he runs off
again laughing his ass off. He wanders off to a
secluded spot, drops his drawers and squats,
still laughing about the trick he pulled.

ANGLE ON: Camper #4, getting ready to take a
dump. Ten feet away, **SOMETHING** is burrowing
toward him, unseen. The earth on the surface is
disturbed. And then the creature is right
underneath Camper #4, its head now visible.

> **CAMPER #3 (O.S.)**
> So the I.R.S. guy begs for mercy but
> the dentist tells him that the
> government showed him no mercy and
> starts to press hard on the screaming
> drill.

With a speed so swift it defies the laws of
physics, the creature forces it way into Camper
#4

Camper #4 screams a blood-curdling scream.

> **CAMPER #3**
> (to the other campers)
> Just ignore him. He's only looking
> for attention.

The other campers laugh, agreeing with the
assessment.

The creature goes up through Camper #4 and comes out his mouth. Camper #4 is vibrating horrifically but still alive. The creature turns around to face Camper #4 with the man's guts hanging from its teeth.

The creature seems to be laughing, taunting Camper #4.

EXT. / INT. HOME / LAB OF EVE AND ADAM REARDON

Re: the news brief that just ended.

> **EVE REARDON**
> Can you believe that? An enemy...a monster we created.

Adam leaves his work station and comes over and stands behind Eve who is focused on hie computers screen, running simulation.

> **ADAM REARDON**
> (pats Eve shoulder to comfort her)
> Not "we," Eve. The enemy is Ronald Davis. Him and his secret lab have unleashed a thing we know little about...certainly not its weaknesses.

> **EVE REARDON**
> (turns to Adam)
> ...if it has any. This could be a totally new life form. Anything we may attack it with: bombs, electricity, radiation, the creatures may use that as a growth factor.

> **ADAM REARDON**
> It may be a new life form but it breathes air and eats food.

47

EVE REARDON

No.

ADAM REARDON

No?

EVE REARDON

It's not breathing our air, it's breathing our waste air: carbon dioxide, nitrogen. And it didn't run out of food, there were fish and plants and fauna.

ADAM REARDON

So why did it migrate?

EVE REARDON

It's diet is the opposite of ours. The original microbes did their job: they ate the radiated waste which caused them to mutate and grow. But they consumed all the available waste. And by cleaning the oceans, photosynthesis started again.

ADAM REARDON

So they ran out of food and they gravitated to land and became mammals. All in the course of what, weeks?

EVE REARDON

Even though it sounds impossible, there's not other explanation that fits the facts.

EXT. BIG ISLAND, HAWAII.   IRONMAN TRIATHLON

The day is all sunshine, with a few puffy white clouds and enough of a breeze to whisk away the sweat.

News media from all over the world have set up trucks, satellite dishes, and stands from which their reporters can get a "bird's eye view." Many of the athletes are being interviewed. Well-placed logos, pictures and banners give great exposure of the sponsors and advertisers of the event. No opportunity is missed to promote a product or service.

The race officials blow their whistles and the contestants do their last minute stretching then line up at the waters edge.

The tension rises parabolically as the media, on-lookers, race teams and security focus all their attention on the triathletes.

And then the whistle blows and the swimmers make a mad dash into the surf.

Quickly, the elite athletes leave the weekend warriors behind.

EIGHT HOURS LATER:

EXT. BIG ISLAND, HAWAII. LATE AFTERNOON

Nearing the finish line of the Ironman Triathlon.

The spectator stands are full in anticipation of seeing the front-runners approach the finish line. There is still a group of ten men vying for the prize.

ANGLE ON: The camera pans the runners and the land around them. On both sides of the path are thick groves of Banana, Hala and Banyan trees, edible plants like coffee and Mountain Apple. Hibiscus, ydrangea, and Bougainvillea. Pheasant, turtle, Raccoon, Toads, Geckos, Asian Mongoose.

There is much movement in the swampy areas on either side of the path. Birds flutter off madly. There is much splashing in the water but mostly hidden from the television cameras that are more focused on the contestants.

As the runners approach the finish line, the lizards run out on to the path. Even as fast as the contestants run, the creatures, whose legs have gotten longer and stronger catch

up to them. The lizards wrap their tails
around the runners' legs and hold them in
place, then slithered up their backsides and
out their throats, with the athlete's
internal organs dangling from their jaws.

In a matter of minutes, all the runners are killed,
their guts strewn all over the road.

The on-lookers, race officials and security are too
stunned to move.

But then the creatures turn their attention to the
stands holding all the people and begin to slowly
edge toward them.

All the people get up at the same time. The stands
tilt over and crash. **PANDEMONIUM.** Cameras aimed at
the creatures go wild as the lizards attack. Then
the cameras go blank.

INT. THE WHITE HOUSE SITUATION ROOM. WASH. D.C. DAY.

The American president, **WALTER DRYSDALE,** a tall man,
who would look more at home on a cattle ranch, is
flanked by his national security team. Their
expressions are grim, mirroring the number and
frequency of disasters being reported.

The military men are in full dress uniforms with
their theaters of combat badges, medals, service
specialties.

They are chomping at the bit to get into a fire
fight. Doesn't matter if it ISIS, al-Quaeda or an
extra-terrestrial.

> **WALTER DRYSDALE**
> Up until now, we have relied on our
> scientific community to determine
> where, when and how to fight this
> menace. While those good men and
> women dedicated themselves to
> finding a solution, their efforts
> have not met with success. I have,
> therefore, turned to our brave men
> and women in the military for their

WALTER DRYSDALE (con't)

advice and support. The plan is now to take the fight to the enemy. We will be going down into the sewer system to attack and wipe out these invaders who believe they can kill our good citizens and get away without retribution.

I will turn the program over to our Seal Team leader, Colonel Wayne Brady.

WAYNE BRADY stands at attention next to the president, rigid and immovable as a Sequoia redwood tree; piercing green eyes, iron jaw. A "take no prisons" expression; eat babies for breakfast.

WAYNE BRADY
(turns to Drysdale)
Thank you for placing your trust in our team, Mr. President.
(turns back to the audience)
We do not intend to let you down and we will attain the objective. We will grind up the chicken-shit bastards and make pocketbooks out of their skins.
(ready to explode)
We will kill all of them, including their ancestors all the way back to the Pleistocene era. Further--.

President Drysdale grabs the Colonel's arm and reins him in.

PRESIDENT DRYSDALE
I think we get the picture, colonel.

COLONEL WAYNE BRADY
Yes, sir. Sorry, sir. It's just that when some one or some thing thinks they can come into our country and try to take over, we get a little bent out of shape.

51

The audience applauds wildly.

President Drysdale holds up Colonel Brady's arm in a premature victory stance.

INT.   HOME / LAB OF ADAM AND EVE REARDON. DAY.

Adam and Eve had been watching the program, then drifted off.

The TV is still on, but the sound is muted. Just a bluish glow.

Music plays, something classical. Bookshelves line the walls, scientific books floor to ceiling.

Eve is curled up in Adam's arms like a Cheshire Cat. We can almost hear her purring.

A glass of Sauvignon Blanc dangles precariously from her hand.

There is a **COMMOTION** outside. **DOGS BARK. BANGING** on the door.

Startled, Adam jumps up, throwing Eve on to the floor.

Adam reaches down to help Eve up. Eve's wine glass spills on the carpet.

                    EVE
          Shit!

                ADAM REARDON
          Jesus, I'm sorry, baby.

The thumping at the door becomes more pronounced.

              ADAM REARDON (con't)
              (yelling)
          Hey, will you hold on a second!
          I'm coming.

EXT. / INT.   THE REARDON'S HOME / LAB. NIGHT

Adam rushes out to see who and what it is. He **OPENS** the door and is face to face with a military major: **RYAN FITZPATRICK** — well over six foot tall, rigid, tough as nails, and his adjutant: **LIEUTENANT KYLE DEAVER** — cut from the same mold.

> **ADAM REARDON (con't)**
> Who are--?

The Major presents his credential.

> **ADAM REARDON (con't)**
> (reads the ID)
> Major Ryan Fitzpatrick.

Major Fitzpatrick looks beyond Adam, scoping out the rest of the house, then turns his attention back to Adam.

> **MAJOR FITZPATRICK**
> Dr. Reardon?

Adam Reardon is suspicious but curious as to what the military wants of him.

> **ADAM REARDON**
> Which Dr. Reardon are you referring to?

Major Fitzpatrick is trying to figure out if Adam is serious or a wise guy.

> **MAJOR FITZPATRICK**
> There' more than one?

> **ADAM REARDON**
> (insulted)
> My wife is also a doctor.

Major Fitzpatrick sees that Adam is serious.

> **MAJOR FITZPATRICK**
> Then I need to speak to both of you.

ADAM REARDON'
Can I ask what it's about?

All this time they have been standing in the doorway, growing more uncomfortable.

MAJOR FITZPATRICK
I'd rather speak to both of you together so I don't have to repeat myself.

ADAM REARDON
(having a brain freeze)
Oh, yeah, right.
(he steps aside)
Please come in.

The major comes in; his adjutant doesn't move. Adam looks at the men, confused. The Major sees it.

MAJOR FITZPATRICK
For security purposes he needs to remain outside.

Adam leads the way into the house.

Eve has been standing at the entrance to the living room, listening.

Adam starts to make the introduction. The major interjects, puts out his hand. His manners are impeccable.

MAJOR FITZPATRICK
Ryan Fitzpatrick, ma'am. Nice to meet you.

Eve is VERY suspicious. It shows on her face and body language.

EVE REARDON
To what do we owe this pleasure? Major?

MAJOR FITZPATRICK
Ryan, please.

EVE REARDON
Ryan it is...the question still
stands.

The major chuckles, impressed by Eve's toughness.

MAJOR FITZPATRICK
You argued against Davies' plan to
introduce engineered microbes to
eat the waste in the ocean.

Eve straightens up, crosses her arms in front of
her chest. She looks like she just swallowed last
year's mayonnaise.

EVE REARDON
It didn't take a rocket scientist
to know he was an accident looking
for a place to happen.

MAJOR FITZPATRICK
But you were the only scientists
to speak up.

EVE REARDON
His plan was insane. The man is a
menace to society, not a savior.
We tried to tell him that. Hell,
we tried to tell the whole word;
they thought our approach was too
slow. Now they've created a monster.

MAJOR FITZPATRICK
Monsters, plural. And now they're
in every country.

ADAM REARDON
That doesn't surprise us.

MAJOR FITZPATRICK
You've been following the news?

ADAM REARDON
Yes, as much as they let out. But
they're holding back. As bad as
things are, there are even worse
things they're not saying.

MAJOR FITZPATRICK

We can't afford a panic.

ADAM REARDON

From the reports we've seen, there already is a panic. Media converge of tens of thousands of people fleeing the costal waterways. Highways backed up for miles. Gas stations robbed of gas and money. Stores broken into for food and supplies. I don't know how it can get much worse.

MAJOR FITZPATRICK

I'm afraid it's already worse.

EVE REARDON

You don't look well, major. Maybe you ought to spit it out.

MAJOR FITZPATRICK

All those people who have fled the inland water ways all over the world are moving in the wrong direction.

ADAM REARDON

Huh?

MAJOR FITZPATRICK

There are no more reptiles or lizards - or what ever the hell they are - in the oceans anymore.

EVE REARDON

Did they die off?

MAJOR FITZPATRICK

If they did, a lot of our troubles would be solved.

ADAM REARDON

If they're not in the oceans and they didn't die off, then where are they?

MAJOR FITZPATRICK

They've entered the sewer systems.

EVE REARDON

If that's true, then they can travel anywhere the pipes go.

MAJOR FITZPATRICK

That's only partially true.

ADAM REARDON
(confused)

I'm not following you.

MAJOR FITZPATRICK

The things are evolving.

EVE REARDON

That's impossible. Evolution happens on a glacial scale, millennia, epoch, eras, ages.

MAJOR FITZPATRICK

What I'm going to tell you next is highly classified.

ADAM REARDON

If you didn't trust us already, you wouldn't be here in the first place.

MAJOR FITZPATRICK

Right. The waste from the nuclear power plants has been seeping into the aquifers for decades.

EVE REARDON

And you did nothing to stop it?

MAJOR FITZPATRICK

I follow orders, not make policy, Ms. Reardon. It may not always lead to the best results but that's what has enabled the army to maintain discipline and win wars.

ADAM REARDON

Why, Major?

                    MAJOR FITZPATRICK
It was due to OPEC. After the 1972
gas embargo, we needed to be less
dependent on others for our energy
supplies.

                    EVE REARDON
Even at the cost to our environment?

                    MAJOR FITZPATRICK
That cost was way down the road.

                    ADAM REARDON
And now it's time to pay the piper.

                    MAJOR FITZPATRICK
What's done is done. We can't redo
the past. What we can, and must do,
is win this war.

                    EVE REARDON
Where do we stand now, major?

                    MAJOR FITZPATRICK
These things don't just swim anymore.
They sprouted legs. At first, just
little fins.

                    ADAM REARDON'
And now?

                    MAJOR FITZPATRICK
Legs like a mammal.

                    EVE REARDON
I know you're telling us the truth,
but as a scientist, I can't accept
what you're saying.

The major takes out some pictures from a shoulder
pouch and hands them to Eve. Adam steps over to
see them as well.

                    EVE REARDON (con't)
My God!

                         58

MAJOR FITZPATRICK

And they're fast.

ADAM REARDON

How fast?

MAJOR FITZPATRICK

They attacked some runners at the
IRONMAN triathlon in Hawaii. The
men tried to out run the things...
and lost.

ADAM REARDON

Christ.

MAJOR FITZPATRICK

He was our first choice but the
Lord preferred not to get involved.
Probably figures we'd have to clean
up our own mess.

EVE REARDON

How many of these things are there?

MAJOR FITZPATRICK
(frustrated)
We don't know exactly, we guess in
the millions, but it might be
billions.

ADAM REARDON

How are you going to kill that many
without carpet bombing the cities.

MAJOR FITZPATRICK

Even that wouldn't work. They go
underground when attacked.

EVE REARDON

What about some gas?

MAJOR FITZPATRICK

First, we don't know what gas, if
any, can kill them. And to try,
we'd have to evacuate every single
city in America...hell, the world.
Where would all those people go?

Eve realizes they haven't offered the major a
seat or drink in all the time he has been there.

                    EVE
              (ready to make amends)
          Shit. I'm sorry major. I've turned
          into a social retard. Can I get
          you anything t eat or drink? A chair
          to sit down?

              MAJOR FITZPATRICK
          No thank you, ma'am. Relaxing doesn't
          help my thinking. I need to stay
          focused.

                ADAM REARDON
          So you've come to us to find the
          answer?

              MAJOR FITZPATRICK
          You knew, at least you were pretty
          Sure, Ronald Davis' methodology was
          dangerous. No one else thought so.
          You must have done simulations.

                 EVE REARDON
          We did simulations on microbes *we*
          engineered.

              MAJOR FITZPATRICK
          Were they similar to Davis'?

                 EVE REARDON
          Close only counts in hand grenades
          and horseshoes. What we did won't
          tell us how to stop those things.

              MAJOR FITZPATRICK
          We've had our best people on this
          and they've gotten nowhere. These
          creatures are going to keep evolving
          and growing in numbers until they
          encompass the planet. They'll eat or
          destroy the entire food chain along
          with all the humans. We don't have
          unlimited time. Hell, we don't have any
          time.

ADAM REARDON
There's no way we--.

MAJOR FITZPATRICK
You need to stop tell me you can't do. No one was closer to Ronald Davies and Paula Yeats' work than you were.

ADAM REARDON
Why not call them?

MAJOR REARDON
They're under house arrest. About to be brought in front of a Senate Sub-committee investigating this pestilence.

EVE REARDON
That sounds biblical, major.

MAJOR FITZPATRICK
I don't think there is any other way to describe what we're facing, ma'am.

EVE REARDON
You're making me feel older than I already feel.

MAJOR FITZPATRICK
I'm sorry, ma...Ms. Reardon.

ADAM REARDON
Can you get us a sample of one of those creatures?

MAJOR REARDON
It's waiting for you outside.

EVE REARDON
You mean the creature or a piece of it.

MAJOR FITZPATRICK.
A piece cut from one, kept on ice.

ADAM REARDON

We'll get right to work on it,
major.

MAJOR FITZPATRICK

I'm going to need a hell of a lot
more commitment than that, sir.

ADAM REARDON

What did you have in mind, major?

MAJOR FITZPATRICK

You'll have at your disposal the
best minds in the military, every
piece of testing equipment, the
super computers. This is a race
against time. Saying 'you'll get
on it' doesn't exactly make me
confident.

EVE REARDON

You said these things are in every
country. Do we know what others
Are doing? Can we establish
communications with them?

MAJOR FITZPATRICK

We can bring you to Lawrence
Livermore Labs. We have satellite
hook ups with every industrialized
nation; also where the super
computers are.

EVE REARDON

We work better from here, familiar
surrounds, quiet, we can think,
something that's not possible at
Livermore where there are probably
hundreds of people from every
scientific discipline demanding
EVE REARDON (con't)
more computer time, trying to
hijack the project, take credit
for any finding, control,
manipulate the work.

MAJOR FITZPATRICK

We can bring in a hookup. You'll
be able to speak to scientists from
the countries we have on line and
pick up others you think can help.

ADAM REARDON

How soon?

MAJOR FITZPATRICK
(a sly smile)
Before you get up tomorrow morning.
(a thoughtful pause)
Can you leave the front and back-
doors open...along with the garage?

ADAM REARDON

Doesn't quite seem equitable. You
work all night while we sleep.

MAJOR FITZPATRICK

The military never sleeps, Mr.
Reardon.

Eve is considering all that has been discussed. Some
things unanswered...important things but jumbled up
with all the details.

Major Reardon sees it.

MAJOR FITZPATRICK
Anything else?

EVE REARDON
What's plan "B"?

MAJOR FITZPATRICK
You don't want to know that.

ADAM REARDON
(pretty sure he knows)
Nuke 'em.

MAJOR FITZPATRICK
(moving toward the
front door)
You didn't hear that from me.

Adam walks the major to the front door and opens it. The major's adjutants hands a small ice chest to Fitzpatrick who hands it to Adam.

Adam and the major shake hands.

The major's men let him pass then walk off behind him.

INT. AL JEZZERA TELEVISION. LONDON, ENGLAND

New anchor Hussein Mustafa sits at his desk sharing the news of the day. He is a dark-complected, handsome man wearing western clothing out of place with his heritage. The man is visibly shaken and has lost his normal cool and calm demeanor.

> **HUSSEIN MUSTAFA**
> Today, we find ourselves a world at war. Not a war against each other, but against a common enemy. Reports are coming in from all over the globe of attacks by what some are calling snakes, other lizards. Whatever the connotation, these things have attacked at will then quickly returned to the sewer systems before any offensive action can be taken. The police and military who have managed to shoot some of the beasts said they are extremely difficult to kill and that as many as a dozen bullets were needed to stop just one of the things.

Hussein tapped his ear piece.

> **HUSSEIN MUSTAFA (con't)**
> We have footage of some of the attacks. Let me warn viewers that these are scenes of horrific violence and gruesome deaths.

The camera pans away from the news anchor and over to the big screen monitor.

ANGLE ON:

IMAGES FROM PARIS, FRANCE show the creatures slithering in and out of the sewers. Some of the things wrap their tails around tourists waking along the Champs-Élysées, dragging them down into the depths. Other lizards attack restaurants, crashing through windows, biting the heads off diners, slithering down their throats to eat their insides. Dozens work together to turn a bus over, leaving the people at the mercy of the creatures. Some of the people try to run away and almost make it to safety when a tail wraps around a leg and drags them into the sewer system.

IMAGES FROM ISTANBUL, TURKEY show hundreds of the creatures attacking a Mosque filled with worshipers. Frantic people run for their lives, only to be met outside by more of the creatures waiting guard beyond the doors, wrapping their legs around the people, slithering up their backsides, coming out their throats, their mouths full of internal organs.

IMAGES FROM ST. PETERSBURG, RUSSIA show the walkways full of pedestrians, tourists, locals, curiosity seekers. And then the attack begins. Thousands of creatures come out of the sewers, dragging people away. Their heads are ripped off, chucked away, then the lizards dive down the throats of its victims to feed on their internal organs.

EXT. / INT. CAPITOL HILL. CONGRESS IN SESSION. DAY.

SENATOR BYRON JENNINGS, southern, aristocratic, mean-spirited, heads the Senate Investigating Committee. He is flanked by eight other members.

Facing the committee are Ronald Davies and Paula Meyers.

The hall is packed with people from many varying backgrounds: fellow scientists, the curiosity seekers and reporters from a dozen different countries, all wanting answers.

The lights from the cameras surrounding Paula and Ronald cause beads of sweat to pop up on their foreheads. Each dabs as the perspirations with handkerchief.

                    SENATOR JENNINGS
          As chair, I do hereby call this
          meeting in session.

Ronald pulls the microphone on his desk closer to him.

                    RONALD DAVIES
          Senator, I would like to--.

                    SENATOR JENNINGS
               (interrupting)
          I'm sure you would, sir. But I am
          in charge and you are here to
          answer questions, not ask them.

                    RONALD DAVIES
          But, sir. If I may, I could narrow
          the scope of this investigation
          for the committee.

                    SENATOR JENNINGS
          I imagine you would, sir. But then
          we might overlook some of the
          charges against you.

Ronald prepares to speak again. Senator Jennings slams down his gavel, driving Ronald back in his seat and away from the table.

                    PAULA MEYERS
               (bursting at the seams)
          I told him, senator. Told him his
          simulations didn't go on far enough.

Ronald Davies reaches out for her arm, gripping the microphone, trying to pull it away from her.

                    RONALD DAVIES
               (to Paula)
          Shut up you fool.

                            66

SENATOR JENNINGS
(bangs his gavel)
I am the one who will decide who
speaks and who doesn't, sir.

RONALD DAVIES
I can assure this committee that--.

SENATOR JENNINGS
You can assure this committee of
nothing, you incompetent pseudo-
scientist.

Ronald begins to speak again. Jennings raises his
gavel, Ronald slinks back into his seat.

SENATOR JENNINGS
You may go on, Ms. Meyers.

PAULA MEYERS
Simulations take days, sometimes
weeks to run their course. We were
dealing with microbes that react
differently under varying circumstances. You need to run each
simulation, first with one microbe
interacting with one substance,
then with all the substances, then
go back and start combining the
substances and seeing how they
effect the microbes. The number of
possible combinations in astronomical. It would takes weeks at
best, years at the outside. Ronald
wanted to be awarded that
contract. He was cutting corners.

RONALD DAVIES
(jumping out of his
seat)
She's obfuscating. She doesn't know--.

SENATOR JENNINGS
(nodding at the marshals)
Officers, I want you to gag and
bind this man. And if he resists,
taser him until he glows in the dark.

RONALD DAVIES
(standing up again)
You can't--.

Just then the marshals, each one as big as a refrigerator, reach Ronald. One grips his arm so tight Ronald winced. The other marshal whispered in the ear.

MARSHALL #1
I suggest you cooperate or you'll find yourself in a psych ward wrapped in a straight jacket and shot so full of Thorizine you'll think you've been abducted by aliens.

RONALD DAVIES
I know my right, you can't--.

Marshal #2 **TASERS** Ronald and he slumps in the arms of Marshall #1. They set him down in his chair, zip tie him. Ronald comes around in a few seconds and starts to speak. Marshall #2 stuffs a sock in Ronald's mouth and tapes it in place.

SENATOR JENNINGS
Ms. Meyers, do you have any suggestions of this committee as to how to deal with this menace?

PAULA MEYERS
Do you mean Ronald or the lizards?

The audience laughs out loud.

Senator Jennings bangs his gavel.

SENATOR JENNINGS
While they are both a menace to society, we will, for now, focus on the lizards.

PAULA MEYERS
Those creatures have evolved beyond
the projections of any computer
program. There were thousands of
elements in the waste and those
compounds, using photosynthesis
mutated into new organisms. Smaller
organisms can create a new
generation of organisms in minutes.
We could be a million generations
beyond what the first waste
materials consisted of.

SENATOR JENNINGS
And so you have no advice for this
committee?

PAULA MEYERS
I would look to God, senator.

TWO DAYS LATER:

EXT. CENTRAL PARK. NY. NY. NIGHT.

The military campaign has turned into a media
circus. TV stations from all over the U.S. and many
foreign countries are filming the event.

Stands have been set up for VIPs. TV trucks with
huge satellite dishes on the roofs stretch to the
sky. Behind the cordoned off area, thousands of
excited citizens have gathered to see the finale of
the war against the lizards.

There are a number of sound bites being heard as the
camera and sound flitter from one to another: all
discussing the situation: Interviews of government
and military officials and video of soldiers exuding
supreme confidence.

MONTAGE:

The event is taking place simultaneously all over
the world in all the largest cities on the various
continents: **LONDON:** British red Berets, **BERLIN:** GSG
9 der Bundespolizei, **TOKYO:** Japanese Special Forces

Group, **MOSCOW**: Spetsnaz special forces, **BEIJING**: elite forces group.

EXT. STREETS OF NEW YORK CITY. NIGHT

There are large tents set up at the entrance of the main sewer systems. The seal team prepares to go into the sewer system. The men are outfitted with state-of-the-art weaponry: laser pointing machine guns, automatic grenade launchers, computer-aimed, gyro-stabilized automatic weapons, C4 plastic with remote detonators, heat and night vision goggles, Dragon skin body armor, voice-activated, secure communications ear pieces. Men and women in uniform **FIELD STRIP LIGHT WEAPONS** with precise movements. Around them, in racks, is an arsenal of advanced personnel artillery.

The soldiers like the feel of the guns, the weight...the authority. Hands move without hesitation. **CLACK. CLACK. CLACK.** They swing laser / computer guided **SMART-GUNS** out of the racks. Using a body brace and **GYRO-STABILIZED SUPPORT ARM; COMPUTER-AIMED, VIDEO TARGETED AUTOMATIC WEAPONS.** A sort of a steadi-cam.

SOLDIER #1 - CAPTAIN KYLE RAVEN - old school, hardcore, a lifer. Been in a dozen war zones. This is old hat to him. Jihadists, suicide bombers, lizards; it's all the same to him.

> **CAPTAIN RAVEN**
> Still nothing from the reptiles?

> **SOLDIER #2**
> Dead on all channels, sir.

SOLDIER #3 comes through wheeling a dolly of hand grenades.

> **SOLDIER #3**
> Clear, please.

Other soldiers step aside, they nod in appreciation for the added weaponry.

TV REPORTER, MIKE HAMMER, jumps aside, trying to stay close enough to film and report the events with his mini-cam. Mike has been around the block a few times, even a grunt himself once, but this is a step up from what he's covered before. He's nervous and it shows.

SOLDIER #4 cruises by with a forklift loaded with **FLAME THROWERS.**

                    SOLDIER #4
                (sarcastically)
          Excuse me ladies.

Mike Hammer steps aside, trying not to get in the way.

                    MIKE HAMMER
                (to Captain Raven)
          What can I do, captain?

                    CAPTAIN RAVEN
          I don't know? What can you do?

                    MIKE HAMMER
          I was a grunt for four years. First
          Gulf War.

                    CAPTAIN RAVEN
                (impressed)
          Can you load weapon?

                    MIKE HAMMER
          Does a bear shit in the woods?

                    CAPTAIN RAVEN
                (laughing)
          Then get to it soldier.

Mike salutes, then goes over to the ammo table where soldiers are loading their weapons.

The soldiers are suiting up in their bulky **COMBAT-ARMOR**...interlocking plates like football padding. They tape their wrists, draw on segmented boots. The sole cleats **CLACK** like hooves.

The men don their **WEB BELTS, PACKS, HARNESSES, HELMETS, COM-SETS.**

Their fingers move methodically over the fastenings. **CLICK. CLICK. CLICK.**

> ### CAPTAIN RAVEN
> Let's move it, girls! On the ready
> line. Let's go, let's go.

The soldiers high five each other, then give a thumbs up to their commanding officers and government officials. Their smiles radiate supreme confidence.

The soldiers proceed to the open manhole cover, just big enough to accommodate the body of the soldiers. Weapons are handed down after the solder climb a ladder to the bottom.

SOLDIER #5, PACKER (big, black, bald) waits his turn.

> ### PACKER
> (to anyone in hearing
> distance)
> I am ready, man. Ready to get
> it on. Check-it-out. I am the
> ultimate shit-kicker...state of the
> badass art. You do not want to
> fuck with me.
> (to Mike Hammer)
> Hey, Hammer, don't worry. Me and my
> squad of ultimate shit-kickers will
> protect you. Check-it-out...

Packer slaps the **SERVO-CANNON** controls on his weapon. Continues trying to impress Hammer.

> ### PACKER (con't)
> Independently targeting, particle-
> beam phalanx. VWAP! Fry half a
> city with this mother. We got
> tactical smart-ammo and phased-
> plasma pulse rifles. We got
> sonic electronic ball-breakers; got
> nukes, got knives, sharp sticks --

Captain Raven grabs Soldier #5 by his battle
harness. His voice is low, but it carries.

CAPTAIN RAVEN
Save it, Packer.

PACKER
Sure, sir.

Hammer nods his thanks to Raven.

All the men are in the sewer system, standing at the
bottom, pipes above and to the sides of them. Murky,
dank water at their feet.

SOLDIER #6 (DAVIS)
It smells like shit down here.

SOLDIER #7 (ROBERTS)
It's probably your shit, Davis. You
fucking live in the bathroom.

CAPTAIN RAVEN
Stow it, Roberts.

Captain Raven reaches back and grabs **SOLDIER #8:
SERGEANT KELLY** (a lifer, grizzled, short, stocky,
veins on his arms and neck popping from lifting
weights). He holds a steady cam, which has night
vision and infrared capabilities.

CAPTAIN RAVEN (con't)
Sergeant Kelly. Take the point. Call
it out when you've got a fix on the
enemy.

Sergeant Kelly is proud to be picked to lead the men
in.

SERGEANT KELLY
Roger that, sir.

Captain Raven turns to Mike Hammer.

CAPTAIN RAVEN
You stick with me.

Mike is happy to do so.

MIKE HAMMER
Ye, sir.

Mike hefts his TV camera on to his shoulder. He is videoing the entire operation.

EXT. CENTRAL PARK, NY. NIGHT

Simultaneously with Mike turning on his camera, the video comes up on a giant screen that can be viewed by all the military, dignitaries and reporters seated in the aluminum stands set up for the event.

BACK TO SCENE:

CAPTAIN RAVEN
(now in combat mode)
We're moving out; maintain distance, silent mode, hand signals.
(whispers to Sergeant Kelly)
Take us in, sergeant.

Kelly nods, maintaining silence as much as possible.

As soon as they move out, some of the taller men bang their helmets on the low hanging pipes.

SEVERAL SOLDIERS
Shit, fuck, ow.

Captain Raven is pissed. He glares at the men who stow it right away and put their game faces on.

The tunnel is poorly lit by overhead bulbs spaced a hundred feet apart. They move on slowly, cautiously.

Captain Raven is right behind Sergeant Kelly. Whispers.

CAPTAIN RAVEN
Anything, sergeant?

Sergeant Kelly's P.O.V: through this night scope.
Bright as a sunny day.

                    SERGEANT KELLY
              (as softly as possible)
          Negative, sir.

No other sounds and movement except for the
soldiers.

Captain Raven turns, points to a SOLDER #9: **CORPORAL
TRAVIS.**

The soldier makes his way silently to the front.

                    CAPTAIN RAVEN
          Travis, watch the rear.

Corporal Travis smiles. Nods. Makes his way to the
rear.

Sergeant Kelly stops abruptly. He looks down at his
feet, to see what the "squishing noise" was when he
set his foot down.

SERGEANT KELLY'S P.O.V.

There are human body parts strewn all over the
ground: heads, guts, legs, feet, arms, but not one
whole person. It's disgusting and several of the men
almost lose their dinners.

Captain Raven turns and gives the men hand signs:
"Be at the ready; we're closing in."

A few emergency lights are still on. Pools of water
cover the floor. Farther down, water drips from
leaky pipes.

Sergeant Kelly moves forward. Taut. Alert.

ANGLE ON: His **STEADY CAM** as he swings it in an arc.

BACK TO SCENE:

Sergeant Kelly studies the locator monitor, looking
down rather than ahead.

75

Their footsteps **ECHO**.

EXT. STREET LEVEL. NIGHT.

All the spectators and military brass and the press watch as the soldiers move through the sewer system below. The big screen set up makes it look like an outdoor movie theater.

BACK TO SCENE:

                    CAPTAIN RAVEN
                (to Sergeant Kelly)
            Heat seeking, sergeant.

                    SERGEANT KELLY
            Roger that, sir.

The tunnel splits. Captain Raven grips the shoulder of Sergeant Kelly. The team stops. Raven thinks for moment, weighing the risks against the rewards.

He has made up his mind.

                    CAPTAIN RAVEN
            Second team, take the left tunnel.
            Corporal Draper, take the point.
            Choose four men. Use your motion
            sensor.

CORPORAL DRAPER, a big ole' Southern Boy, chomping at the bit, starts to reply in a loud voice. Captain Raven winces. Draper begins again, whispering.

                    CORPORAL DRAPER
            Roger that, sir.

Corporal Draper points to four men who come forward.

The four soldiers, with Draper at the point, start off down the second tunnel.

Captain Raven nods at Draper who nods back. Their expressions do not exude confidence.

Sergeant Kelly swings his motion detector from side to side. He adjusts the "gain." It remains silent.

The amount of water at their feet increase. Captain Raven shines a light. It's **BLOOD**, not water. Human body parts float in the effulgence.

SOLDIER #10
Looks like a frat party turned ugly.

Nobody laughs.

There is blood curdling scream that stops the shoulders in their tracks.

Captain Raven calls out to a soldier at the rear.

CAPTAIN RAVEN
Who was that Edwards?

Private Edwards is a small soldier, carrying a radio almost as big as he is.

PRIVATE EDWARDS
(spooked
Travis, sir. Should I go back and check?

CAPTAIN RAVEN
Negative soldier. Call it in. Have someone from up top go in.

PRIVATE EDWARDS
(relieved it's not him)
Roger that, sir.

Captain Raven touches a splash of color on the wall{ **BLOOD**.

Sergeant Kelly's tracker **BEEPS**.

CAPTAIN RAVEN
(in highest alert mode)
Where, sergeant?

Sergeant Kelly slowly wheels around, the **PINGING GROWS MORE FREQUENT AND LOUDER.**

> **SERGEANT KELLY**
> (the bearer of bad
> tidings)
> All around us, sir.

The men come to a crossroads. There are now four directions they can be attacked from.

There are screams coming from the tunnel to the left of them.

All the men freeze.

A soldier is running toward them: one of the men under Corporal Banner.

The soldier almost reaches Captain Raven when he is suddenly jerked off his feet and dragged back into the darkness.

His screams continue, but only for a little while.

The soldiers all bring their guns to shoulder height.

> **CAPTAIN RAVEN**
> (with hand motions)
> Hold your fire!
> (turns to his radio
> man)
> Private Edwards. See if you can raise
> second squad.

Sergeant Kelly continues to sweep the area.

TIGHT ON HIS MONITOR as the pinging shows...nothing.

BACK TO SCENE:

Sergeant Kelly looks at Captain Raven. His expression is grim.

> **PRIVATE EDWARDS**
> Nobody's home, sir.

CAPTAIN RAVEN
They're out there, sergeant. They
just don't throw off any body heat.
Go in night vision.

Captain Raven turns to SOLDIER #9: **CORPORAL RAPHAEL**

CAPTAIN RAVEN (con't)
Raphael, take one man and scout
ahead.

Corporal Raphael is an inner city kid. Small, wiry,
tough at nails.

CORPORAL RAPHAEL
(smiling at the captain)
Roger that sir.

Raphael and another soldier take off at double time.

Only a minute later, horrific screams come from the
tunnel just a few hundred yards in front of them.

Corporal Raphael is running / **STAGGERING** toward
them.

Captain Raven is about to help him when a serpent
comes out of the corporal's mouth with Raphael's
guts dangling from its jaws.

All the men under Captain Raven raise their weapons
but don't fire.

CAPTAIN RAVEN
Shoot, goddamn it!

SOLDIER #6
But Corporal Raphael, sir!

CAPTAIN RAVEN
He's dead, soldier. Now fire!

The squad opens fire on the serpent, also hitting
Raphael.

The serpent is hit a dozen times, wounded, but still able to slither back into the darkness.

The wind howls through the tunnel.

Captain Raven looks to Sergeant Kelly with a "What was that?" look.

> **SERGEANT KELLY**
> Those things must have opened a manhole cover somewhere down the line, it's fresh air coming in.

> **SOLDIER #8**
> Maybe they're retreating, sir?

> **SERGEANT KELLY**
> Or maybe more of their friends are coming to reinforce them?

All the lights in the tunnel go out.

Captain Raven throws a look at Sergeant Kelly: "Did they do that?"

The soldiers switch on their pack lights and the beams illuminate a scene of despair worse than they could have imagined.

Right in front of them: the dead, partial bodies of victims of the serpents are piled up almost as high as the ceiling in the tunnel.

The Captain considers what they face.

He turns to Sergeant Kelly.

> **CAPTAIN RAVEN**
> (whispers)
> These things are evolving, sergeant. They're beginning to think like us: create a scene so horrific that we'll give up the pursuit.

> **SERGEANT KELLY**
> I thought they were just snakes, or lizards or something.

CAPTAIN RAVEN

They were...once, not now. They
went from gills to lungs, flippers
to feet, now reptilian brains to
homo sapiens.

SOLDIER #7

Do we retreat or go on sir?

Captain Raven is about to speak when they are
startled by a loud **BEEP**. Sergeant Kelly is intent on
his motion tracker, aimed back toward the rear.
BEEP. BEEP.

SERGEANT KELLY

Behind us.

He gestures at the corridor they just passed
through.

CAPTAIN RAVEN

One of us?

SERGEANT KELLY

Negative, sir. Dozens of signatures.

CAPTAIN RAVEN
(turns to his radio man)
Private Edwards. See if you can
raise our people.

Sergeant Kelly rotates his scanner 360^0

SERGEANT KELLY
(grim expression)
Negative, sir.

PRIVATE EDWARDS
(to Captain Raven)
No response, sir.

CAPTAIN RAVEN

Fuck these lizards. Takes us in
Sergeant.

He turns back to a solider: **CORPORAL FLYNN**, carrying a flame thrower. The soldier is a monster, muscled, thick.

                        CAPTAIN RAVEN
          Corporal Flynn. Take the point. Fry
          the fuckers.

                        CORPORAL FLYNN
                (a shit-eating grin)
          With pleasure, sir.

Corporal Flynn takes out a striker and **SNAPS** it in front of the nozzle. The gun shoot out a small flame: just a test to see if it's not clogged.

The squad heads toward the source of the signal, the others following.

Sergeant Kelly's tracker is reading out more rapidly.

Radio man, Private Edwards, hangs back. Then realizes there is nothing behind him but darkness. He catches up to the group.

Sergeant Kelly is scanning, gaze intense. The other soldiers grip their weapons tightly.

They come to a "T"

                        CAPTAIN RAVEN
          Which way?

SOLDIER #11 trips over an extruded pipe.

The squad points their guns, fingers on the triggers, in the direction of the man who fucked up.

When they realize it was one of their own, they glare at Soldier #11 who gives a sheepish grin.

Sergeant Kelly's tracker beeps steadily. The beeps from the two tunnels merge, becoming a solid tone.

                        SERGEANT KELLY
        Movement!

                    CAPTAIN RAVEN
          Position?

                    SERGEANT KELLY
          Can't lock on.

                    CAPTAIN RAVEN
                (with an edge)
          Talk to me, sergeant.

                    SERGEANT KELLY
          Uh, seems to be in front and behind.

                    CAPTAIN RAVEN
                (into his shoulder radio)
          Corporal Barnes, report.

                    CORPORAL BARNS (voice)
          We can't see anything back here, sir.

Captain Raven thinking to himself.

                    CAPTAIN RAVEN
          What's going here?

The Captain senses it coming, like a wave at night.
Dark, terrifying and inevitable.

The creatures rush out of the dark heading right for
the squad.

Private Flynn laughs manically, frying the
lizards, driving them back...for moment.

The other soldiers sense a victory and open up with
their weapons on full automatic. A stream of tracers
fill the tunnel.

An orgy of purging fire. A line of lizards go down,
but there are an endless number behind those.

                    CORPORAL FLYNN
          Eat shit and die, you fuckers!

A tail whips into frame, pulling the flame thrower right out of Flynn's hands.

He looks down. It seems like a magic trick.

The lizards race forward, actually climbing on top of the slower ones, until they overwhelm the soldiers on the front line and begin to consume them.

A SCREAM.

Captain Raven whirls, uncertain.

> **CAPTAIN RAVEN**
> Edwards? Riley? Sound off!

> **SOLDIER #7**
> (freaked)
> We're getting screwed! We're
> gonna die in here!

SOLDIER #12 nearby, is laying down a horrendous field of fire. He pivots, firing mechanically in controlled bursts. Scoring points in his own private video game.

Private Flynn unslings his AR-16 and begins firing hundred round clips at twelve seconds each.

Sergeant Kelly pushes a dead solider aside, leaps in after Captain Raven and drags him back, massive gear and all. He sees a lizard lunge toward him, its mouth opening wide, showing two sets of carnivore teeth.

Sergeant Kelly frees one hand to raises his 12-gauge. A lizard steps on the dead bodies of his family, its hideous mouth opening. Sergeant Kelly jams his **SHOTGUN MUZZLE** between its jaws and pulls the trigger! **BLAM**! The creature is flung backward, its shattered head spouting blood.

EXT. STREET LEVEL. NIGHT

The action below is shown on the big screen set up just outside the main entrance of the sewer system.

84

The military brass, elected officials, news media, are stunned, uncertain what their eyes tell them. Some of the guests begin throwing up as the lizards crawl into the mouths and backsides of the soldiers. Human intestines, like stringy spaghetti, litter the ground and hang from the jaws of the beats.

GENERAL BRADLEY
Pull you team-. Static. Captain I order you to-. Static.

BACK TO SCENE.

The lizards retreat.

There is a momentary lull in the action. Captain Raven and his men regroup. Did they win the battle or have they been lulled into a sense of victory?

CAPTAIN RAVEN
Go to infrared. Looks sharp people!

The squad members snap down their imaging-goggles.

SERGEANT KELLY
(to Captain Raven)
Multiple signals. All round.
Closing.

SOLDIER #11: PRIVATE ROSA RODRIQUEZ (a female) steps forward, bumping other soldiers out of the way. She is carrying a pulse rifle that puts out a laser beam at four thousand degrees Fahrenheit.

Captain Raven is pissed.

CAPTAIN RAVEN
I thought I told you, private, to--.

Rosa cuts him off at the knees.

PRIVATE RODRIGUEZ
With all due respect, captain, fuck that. My homies are dying and I'm not going to sit on my ass and kiss

my rosary beads.

The lizards charge from out of the dark.

Private Rodriguez pushes Raven aside and opens up
with the laser gun vaporizing the creatures, turning
the tide of the battle.

The attack slows down.

                    CAPTAIN RAVEN
          Cease firing. I ordered hold fire,
          Damm it!

PRIVATE RODRIGUEZ rips off her headset. She is
riveted to her **TARGETING SCREEN.**

ANGLE ON: SERGEANT KELLY'S MONITOR.

They're still out there. **SQUIGGLES** fill his screen.

BACK TO SCENE:

Instead of retreating, Private Rodriquez steps
forward. She is moving like a ferret, firing
pivoting, almost a dance. Better than sex for her.

The serpents **SQUEAL** like pigs when they are hit.

Rodriguez laughs.

                   PRIVATE RODRIGUEZ
          Eat shit and die, motherfuckers.

A lizard **SCREECHES** from the darkness.

Private Rodriguez presses the trigger again.

NOTHING.

She looks at the gun, then looks up at the baby face
of the creature just inches away from hers. It tears
her head off with the ease of opening a can of soda.

As Captain Raven, Sergeant Kelly and the few
soldiers still alive watch, the serpent slithers
down Private Rodriquez's neck and comes out her

backside, with **ROSA'S GUTS** trailing behind the creature like confetti in a ticker-tape parade.

Captain Raven and Sergeant Kelly and two other soldiers open fire, killing the creature, but not before it has killed Private Rodriguez.

Private Flynn IS **FREAKED.**

> ### PRIVATE FLYNN
> Let's get the fuck out of here, sir.

> ### SERGEANT KELLY
> (the bearer of bad
> tiding)
> There is no "out of here" soldier.
> We're surrounded. We make a stand.
> Take as many of them with us as we
> can.

Captain Raven looks over at Sergeant Kelly's monitor.

ANGLE ON: Sergeant Kelly's screen.

The screen is filled up entirely with the images of serpents.

BACK TO SCENE:

Captain Raven and Sergeant Kelly are down to their AR-16s. Not much against a superior enemy.

They **OPEN UP** simultaneously, lighting up the tunnel like welders' arcs as the serpents attack.

EXT. TOPSIDE. STREET OF NYC. NIGHT

General Bradley watches, horrified as his picked team of top men are sliced and diced. If these men can't stop the lizards, then they are all doomed.

General Bradley's monitor shows the number of life signs of his men in the tunnel falling, until only three or four are left.

BACK TO SCENE:

A call comes in to Captain Raven.

> **GENERAL HARPER (voice)**
> Get your men. **STATIC**, captain!
> (static)
> Get
> (garbled)
> squad to
> (garbled)

General Harper's voice breaks up completely.

INT. HOME / LAB OF ADAM AND EVE REARDON. MORNING

It is early morning. Robins begin chirping.

ANGLE ON:

The Bedroom of Adam and Eve Reardon. Walls covered with art, a thick, white, down comforter on the bed along with a load of brightly colored pillows: woman's touches. Adam tries to slip out of bed without disturbing Eve.

Full of sleep, Eve reaches out for Adam's arm. She touches it but her hand slips off, indicating how weak she is. The cancer is spreading, taking over her body.

> **EVE REARDON**
> Where are you going? It's what...
> (she looks at the clock)
> Five thirty?

> **ADAM REARDON**
> (sits down on the bed,
> takes Eve's weak hand).
> Major Fitzpatrick's men have been working all night. I want to see what they've done.

> **EVE REARDON**
> (trying to sit up)
> I'll be down in a minute.

ADAM REARDON
(gently pushes her
back down)
Why don't you go back to sleep and
I'll come up and get you once I
check out the system.

EVE REARDON
(no fight left in her)
Okay, but don't let me sleep the
day away.

Adam walks down stairs and see what the military has
set up. He is stunned by the degree of speed and the
amount of technology sitting in his dining room.

There are six computer screens in the living room.
Each has a key board as well as voice activated
mics.

As Adam walks by each screen, a face and a voice
say hello in different languages: French,
Portuguese, Italian, Russian, German and Japanese.

The audio / video is two way. Adam sees and is seen
as he walks by.

JAPANESE SCIENTIST
Mr. Reardon, Dr. Toriata here. We
are e-mailing you the results of
experiments we have conducted using
varying degrees of electricity. It
seems the creatures are susceptible
to shock but we have not determined
how many volts it will take to kill
them and how to delver the
electricity in away that affects
only the creatures.

ADAM REARDON
Good work, doctor. Have you been
supplied one of the creatures to
experimented on?

DR. TORIATA
So far, only samples cut from the
lizards.

ADAM REARDON

I'll see to it that you get a live
one to work on.

DR. TORIATA

That would be wonderful, doctor.
Please keep us informed.

The Russian scientist, **DOCTOR KOFSKY**, calls out
Adam's name.

ADAM REARDON

Yes, doctor?

DOCTOR KOFSKY

We concluded that since they quickly
gravitate back to the sewers after
an attack, that they may be light
sensitive. We are experimenting with
bright lights. So far it seems to be
having an effect but only at high
lums.

ADAM REARDON
          (waiting for the other
          shoe to drop)
How high, doctor?

DOCTOR KOFSKY

Hum, one million cable power, the
same as a lighthouse.

ADAM REARDON
          (little patience,
          sarcastic)
Oh, is that all.

DOCTOR KOFSKY

We are working with our military
how to get the light into the sewer
systems.

ADAM REARDON
          (unconvinced, a dead
          end)
That sounds promising, doctor, Can

you please keep me updated?

DOCTOR KOFSKY
Yes, of course. We must not allow
our ideological differences to
interfere with our work here.

A third screen comes alive. A German Scientist, **DR. CONRAD DUEREN** calls out Adam's name.

ADAM REARDON
(his pace quickening
with each call)
Yes, doctor?

DOCTOR DUEREN
We have one of the creatures alive
and in our possession.

ADAM REARDON
Excellent! What test have you
performed?

DOCTOR DUEREN
We first had to determine its sex.

ADAM REARDON
(anxious)
And...?

DOCTOR DUEREN
Female.

ADAM REARDON
That's--.

DOCTOR DUEREN
A pregnant female.

ADAM REARDON
Wow! What have you been about to
determine?

DOCTOR DUEREN
The gestation period is very short.

ADAM REARDON

(wary)
How short?

DOCTOR DUEREN
Days...maybe hours. We're still
working on that.

ADAM REARDON
How are you thinking of applying that
knowledge to our problem.

DOCTOR DUEREN
At first, we thought of sterilizing
the creature, then sending it back
into the pack. But...

ADAM REARDON
But...doctor?

DOCTOR DUEREN
We estimate that based on the density
of the population just in one of our
cities, it would take several hundred
thousand sterile females to slow,
then stop, then reverse their numbers.

ADAM REARDON
And just how long do you think it
might take to eliminate them?

DOCTOR DUEREN
Um, maybe eighty years.

ADAM REARDON
Well, I guess we can eliminate that
line of thinking.
(hesitation)
Don't lose faith, doctor. Remember
Thomas Edison failed a thousand
times before he invented the light
bulb.

DOCTOR DUEREN
Unfortunately, we do not have the
luxury of that many failures.

Adam's cell phone rings.

He walks away to remove himself from the noise before answering.

                    MAJOR FITZPATRICK
          How are you making out, doctor? Any
          hot leads?

                    ADAM REARDON
          Lots of leads, but none of them
          immediately implemental or effective,
          so far.

                    MAJOR FITZPATRICK
          I assume you're speaking of the ones
          coming in from foreign countries?

                    ADAM REARDON
          Correct.

                    MAJOR FITZPATRICK
          What about your theory?

                    ADAM REARDON
          I've been so busy coordinating those
          coming in from abroad that I haven't
          had time to pursue mine.

                    MAJOR FITZPATRICK
          Make time, doctor. We hooked you up
          to those other countries and
          scientists for diplomatic reasons.
          We don't think they're on the right
          track or we wouldn't have gone to
          the trouble to hook you into the
          super computers.

                    ADAM REARDON
          I understand, major. I'll get my ass
          in gear.

                    MAJOR FITZPATRICK
          Then hang up and get to it, sir.
          You're burnin' daylight.

Adam ends the call, laughs at the way the military
phrases things. Adam turns when he hears Eve coming

93

down the stairs. She is a little shaky; holding on the railing for security.

> **EVE REARDON**
> (toughing it out)
> What's so amusing? I thought we were in a life and death struggle for the survival of humanity.

> **ADAM REARDON**
> Laughter it the elixir of life. I thought it might open up some pathways between my axons and dendrites and lead to an epiphany.

Eve walks into the dinning room, laughing.

> **EVE REARDON**
> And did it?

> **ADAM REARDON**
> The epiphany I was waiting for was you: a new line of thinking, a different set of eyes to look at data.

Eve laughs, she has the look of a Leprechaun hold great secrets.

> **ADAM REARDON**
> (certain)
> You do have something.

> **EVE REARDON**
> (a little smirk of
> confidence)
> Maybe. Microbes are released into an ocean permeated with human and radioactive waste and almost devoid of oxygen. Correct?

> **ADAM REARDON**
> So far.

EVE REARDON

The uppermost layer of the oceans
is bathed in sunlight during
the daytime. This bright ocean layer
is called the euphotic zone. The
depth of this zone depends on the
clarity or murkiness of the water.
In clear water, the euphotic zone
can be quite deep; in murky water,
it can be only 50 feet deep. On
average, before the waste dumping,
it used to extend to about 660 feet.
Under ideal circumstances, photosynthesis would produce 10 ppm
oxygen.

ADAM REARDON

I do know all this, you are aware?

EVE REARDON

Bear with me, Mr. Impatient.

ADAM REARDON
        (flustered)
Right.

EVE REARDON

Photosynthesis is a process in which
sunlight and carbon dioxide gas are
converted into food and oxygen.
Photosynthesis in the oceans creates
approximately 50% of the Earth's
gaseous oxygen. Most of the oxygen
is produced by phytoplankton, the
first link in the food chain in the
oceans. Because of this food source,
many animals also live in this zone.
In fact, most of the life in the
ocean is found in this zone.

ADAM REARDON

You're still in biology 101.

EVE REARDON

I'll try to speed things up.
We know there was enough radioactive waste in the oceans to
trigger a massive mutation, not
only a mutation on the physical
level, but also on the reproductive
level. Speeding up the cycle
hundreds, if not thousands, of
times.

ADAM REARDON

Which still leaves us with the
problem of getting rid of them.

EVE REARDON

(flustered)
I was getting to that.

ADAM REARDON

(recalcitrant)
Sorry.

EVE REARDON

The key to that is to learn why the
creatures left the oceans: a
beneficial environment, to one
with more competition and greater
risks. If we can figure that out,
we may learn what their weaknesses
are.

Eve Reardon walks over to her computer, fires it up.

EVE REARDON (con't)

We need to approach this
scientifically, not emotionally.

ADAM REARDON

Even though we're dealing with a
life form that have more in common
with extra-terrestrials than they
do with us.

EVE REARDON

(assuredly)
They are a mirror image of us.

ADAM REARDON

In what way?

EVE REARDON

We take in oxygen and expel carbon
dioxide. They take in carbon
dioxide and expel waste gases.

Adam leans in close to Eve and sniffs her hair. It's
like an aphrodisiac to him. He leans in closer and
kisses the back of her neck.

EVE REARDON (con't)
(she turns around)
Adam! This is serious!

ADAM REARDON
(going in for a kiss)
This is serious too!

EVE REARDON
(pushing him away
playfully)
We have important work to do and
you're only focused on your
primitive needs.

ADAM REARDON

Okay, you win the battle, but not
the war.

EVE REARDON

Fine, Let's start with what we know.
What was the make up of the oceans
before Davis dropped in his
microbes?

ADAM REARDON

Oxygen levels were zero parts per
million. Unprecedented historically.
Not technically zero, just too low
to measure.

EVE REARDON

What's happened since that quack
spilled his microbes into the ocean?

ADAM REARDON

The microbes ate the waste, which we have to assume contained radioactive materials. The microbes started to evolve at a parabolic rates, turning those microbes into creatures twenty feet long.

EVE REARDON

What else?

ADAM REARDON

The waste is consumed, the sun starts to hit the plankton again, producing oxygen.

EVE REARDON

And then, just as suddenly, the amphibians leave the oceans, become land-dwelling mammals that continue to evolve - replacing gills with lungs and fins with legs, and start feasting on humans.

ADAM REARDON

Why do they leave the oceans? They were still evolving, they had enough to eat.

EVE REARDON

The waste in the oceans was laced with poisons. It must have affected them somehow, maybe slowed their evolution, maybe caused mutations.

ADAM REARDON

Poisons to us, a feast for them. And maybe it sped up their evolution, not slowed it down.

EVE REARDON
(touches Adam's arm)
Damn, I didn't think of that.

ADAM REARDON

Even if it affected some of them,
how would that information reach
all of them spread out over
thousands of miles of oceans on
three continents.

EVE REARDON

Those creatures all evolved from
the exact same organism. They somehow evolved with a common brain.
What one learns is instantly
transmitted to the rest.

ADAM REARDON

That's a huge leap of science.

EVE REARDON

Maybe, but it's the only theory
that fits what has occurred.

ADAM REARDON

So they found another, ready source,
and started eating people.

EVE REARDON

They weren't eating the people.

ADAM REARDON

Huh?

EVE REARDON

All the bodies were recovered. The
creatures only ate the internal
organs of the people.

ADAM REARDON

I don't follow you?

EVE REARDON

The waste, Adam, they were eating
the waste.

ADAM REARDON

And then they ran out.

ADAM REARDON

They hadn't run out of people but
they still came ashore.

EVE REARDON

So what else changed?

ADAM REARDON

Well, the oxygen content.

EVE REARDON

(a light in her brain
goes off)
That's it! When the creatures
consumed enough of the waste in the
ocean, photosynthesis began again,
the oxygen levels went up ten or
twenty fold.

ADAM REARDON

Damn. Eve. You are one smart chick.
Now do I get a kiss?

EVE REARDON

Okay. But just one.

Adam bends down to kiss Eve. He slips his hand
inside her blouse. Eve jumps back.

EVE REARDON

I thought you said a kiss?

ADAM REARDON

I always cup a breast while I'm
kissing a woman.

EVE REARDON

Back to work.

ADAM REARDON

I was working.

EVE REARDON

Back to the lizards.

                    ADAM REARDON
                         (exasperated)
          Right.

                    EVE REARDON
          You're the one who got it right
          about the oxygen. What else?

                    ADAM REARDON
          The oxygen levels on land are down
          ten percent from the industrial
          revolution, trees are cut, the rain
          forest destroyed, therefore less
          oxygen produced by land-based
          sources.

                    EVE REARDON
          So...?

                    ADAM REARDON
          The creatures can only live in a
          narrow range of oxygen and carbon
          dioxide.
                         (stops to smile)
          Can I get another kiss?

                    EVE REARDON
          Okay, but just one.

Adam leans in, kisses her, slips his hand in her
blouse. She reaches back, unhooks her bra, allowing
Adam to touch her bare breast. He is so surprised he
pulls his hand out. Eve buttons up her blouse.
Adam looks forlorn. Like a kid whose favorite toy
was taken away.

Adam reaches in again for her breast. She clutches
his hand.

                    ADAM REARDON
          I wasn't done.

                    EVE REARDON
          After we solve the puzzle.

ADAM REARDON

The problem is solved.

EVE REARDON

No so fast, mister. The details;
the devil's in the details.

ADAM REARDON

The creatures can't live in an
environment where there is too
much oxygen. They thrive in a high
CO_2 environment. CO_2 levels are
now over 400 parts per million,
unprecedented in human history.

He tries to reach in to Eve, she anticipates his
move, blocking his hand.

EVE REARDON

And just how do we raise the oxygen
Levels high enough to kill the paleo
creatures?

ADAM REARDON

Johnny Appleseed.

EVE REARDON
(a step behind Adam)
Huh?

ADAM REARDON

Trees. We plant trees. Trees and
flowers.

EVE REARDON

But how do we accomplish that when
people are cutting trees, not
planting them?

ADAM REARDON
(the light goes off
in his head)
There is a second way, a
complementary way we can get
the world to produce less air
pollution.

EVE REARDON

Adam, you are amazing!

ADAM REARDON

And you thought I was just a pretty face.

EVE REARDON

I could kiss you!

ADAM REARDON

We already reached that plateau. I was thinking...

EVE REARDON

Let's stay on message, mister. The world's needs come before your needs.

ADAM REARDON
(frustrated)
Right. We're at 410 ppm CO_2 right now. We'll need to get down to 350 ppm in order to replicate the percentage change in the gases in the ocean.

EVE REARDON

And how do we accomplish that in a narrow time frame?

ADAM REARDON

We get all the industrialized nations to shut down their factories, mining, and transpiration for a week

EVE REARDON
(challenging)
Unprecedented.

ADAM REARDON
(a sly smile, ready
with an answer)
Not so. China shut down every factory for two weeks prior to the Olympics there in 2008.

                    EVE REARDON
But now it's going to take all the
nations to do the same thing.
There's never been that level of
cooperation.
                    (spins to face Adam)
Just how is that going to happen?

Adam cups her chin in his hand and lifts it to look
him in the eye.

                    ADAM REARDON
You're going to do it.

TIME BREAK.

INT. THE LAB / HOME OF ADAM AND EVE REARDON

Both Adam and Eve are focused on their separate
computer screens, concentrating intensely.
Adam walks over to Eve's station and places his arm
gently around her shoulder.

                    ADAM REARDON
                    (absentmindedly strokes
                    her hair)
Okay, what do we know and what don't
we know?

                    EVE REARDON
I know that I'll never get any work
done if you keep trying to seduce me.

                    ADAM REARDON
I do some of my best work after I'm
satiated.

                    EVE REARDON
And I do no work trying to satiate
you.

                    ADAM REARDON
                    (gives up...temporarily)
Right now, the trees and plants
and flowers absorb 450 gigatons of
carbon dioxide a year and expel
approximately the same amount of

oxygen. All we need is to increase
that level to 540 gigatons.

EVE REARDON

In plain English, Adam. Bottom line.

ADAM REARDON

Right now, there are 3 trillion
trees. That's 420 trees per person.
All we need is to increase that by
40 trees per person. Less if we
consider plants and flowers as well.

EVE REARDON

And just how are you going to
convince the majority of the world
to listen to you, stop their lives
and do something that won't earn
them a dime?

ADAM REARDON

I'm not.

EVE REARDON

Then why are we having this
conversation.

ADAM REARDON

Because you're going to talk to
The United Nations.

EVE REARDON

And you arranged this?

ADAM REARDON

No, synic. I went ot Major
Fitzpatrick who went to Colonel
Brady who went to President
Walter Drysdale It's all set.
You'll convince them it's the right
thing; to set aside their personal
obligations and join the rest of
the world in this effort.

EVE REARDON

Why me?

ADAM REARDON

You're a woman. Men listen to a
message coming from a woman better
than when it comes from another guy.

EVE REARDON

And also because I'm dying.

ADAM REARDON

Aren't you the one who said you
wanted to make a difference before
you?

EVE REARDON

Shit. You hand me the greatest gift
I could give to mankind and I whine
like a spoiled brat.

ADAM REARDON

I'm glad you said that, not me.

EVE REARDON

So, is that all?

ADAM REARDON

There is one more thing but I'm
saving that for dessert.

INT. UNITED NATIONS, NEW YORK CITY. DAY

The assembly hall to packed full with
members on the ground floor and guests and
visitors on the second level.

Nigel Cameron huddles with several people
while the buzzing from the audience gets
louder.

Nigel notices the time, breaks off his
discussion and walks to the podium.

It requires little effort to get the
attention of the audience.

SIR NIGEL CAMERON

We all know why we are gathered here today. And that is to put an end to the scourge facing all nations. When we sought out solutions to cleaning the oceans, we went with the most expedient, but least proven methodology. And we have paid a great price of that. We are determined not to make that mistake again. This time we have enlisted the aide of a team of scientists who have experimented, made their calculations and ran many simulations to determine its efficacy. They, and we who have reviewed that work with the finest minds on the planet, feel confident that this course of action is the one and only one that will, once and for all, end the nightmare. Yet for some, who want an instant solution to this problem, they will be disappointed. Yet we will need the cooperation of all to institute this plan and so we must set aside our differences and unite here and now. And if we are successful, this effort may establish a new level of cooperation that may spill over into social programs to care of the indigent, and scientific cooperation that may leas to solving problems that once looked intractable. I would now like to present Ms. Eve Reardon.

Nigel turns to face Eve, smiles and holds out a hand.

Eve sits nervously next to Adam. She is struggling to find the courage to face such an august body. Adam nudges her and she gets to her feet.

Nigel sees her hesitation, walks over and escorts her to the podium.

The audience sympathizes with Eve and applauds her efforts.

Eve adjusts the mic to her level.

> **EVE REARDON.**
> Thank you all for your warm
> welcome and generous support.
> What I, we...

Eve turns to Adam and nods, recognizing his contributions to their research.

> **EVE REARDON (con't)**
> ...my husband and I propose today
> will require the cooperation of
> all nations. We hsve found that
> these creatures can only live in a
> very narrow band of conditions.
> When the oxygen in the oceans rose
> from 1 part per million to 8-9
> parts per million, the lizards
> fled inland. Therefore, we only
> have to raise oxygen levels by 2%
> in order for the lizards to die
> off. And that can be accomplished
> by every adult in the world
> planting one tree or by building a
> small garden or expanding the one
> they already have. In a matter of
> a few days to a few weeks, the
> creatures will die off.

There are murmurs from the audience
questioning the veracity of Eve's program;
doubts about its implementation.
Eve recognizes it, and goes to plan 'B."

> **EVE REARDON (con't)**
> There is an additional action we
> can take to speed up the process
> or to be less reliant on the
> planting. And that is to lower CO_2
> levels.

This causes even more skepticism in the form of muted conversations.

Eve speeds up her presentation.

> **EVE REARDON (con't)**
> Before you dismiss such a plan, you should know that this has already been done on a small scale.

The inertia of the audience picks up in the form of people edging forward in the seat so as not to miss what comes next.

> **EVE REARDON (con't)**
> In China, in preparation for the 2008 Olympics, Beijing ordered the closing of all factories within ten miles of the event centers to shut down for two weeks. If we can extrapolate this effort to include all industrialized nations, then the carbon dioxide levels will decrease and require half as many trees need be planted, speeding up the process.

There are murmurs of agreement from the audience.

> **EVE REARDON (con't)**
> If China could accomplish this for the Olympics, surely the other nations of the world can emulate that effort. We can defeat these creatures in a matter of weeks, but we may be able to continue some of the good begun here. Can't we place restrictions on the most polluting of factories, give incentives for instituting measures to clean their emissions, pay farmers to continue planting trees, provide economic aide to Brazil to preserve the rain forest.

EVE REARDON (con't)

We can do this; we must do this if
we are to survive, because the
lizards were a warning, worse may
yet come if we do not work
together in this common cause. And
we will once again take back
control of our planet. But the
benefits of these plants and trees
will go on far after the creatures
are gone. Those plants and trees
will take CO_2 out of the atmosphere
and turn it into even more oxygen.
The CO_2 in the atmosphere will fall
below 400 ppm, possibly even below
300 ppm, halting the warming of
the Earth, returning to a point
not seen since the beginning of
the industrial revolution. For the
first time in centuries, all the
nations of the earth will bond in
a common cause and everyone, from
the biggest corporations who have
had to pay carbon taxes, to the
farmers in Africa who will see
their crop yields improve, will
benefit from this common cause.

Eve steps back from the podium.

She is shaky on her feet. Nigel rushes over to catch
her as she falls in his arms.

INT. HOME / LAB OF ADAM AND EVE REARDON. DAY

Even a day later, Eve is exhausted. She is lying on
the couch, feeling sorry for herself. Maybe she
aided in saving the world but she can't save
herself. Her time is drawing near and there is
nothing standing between her and her maker. She is
resigned, and in her resignation she is slipping
away even faster.

Eve sits on the couch, her feet curled up under her. She is on the verge of tears, but tuffs it out, not willing to let Adam see her weakness.

Adam comes into the room smiling, whistling.

Eve is visibly angry that he is so insensitive to her illness.

EVE REARDON
(thick with sarcasm)
Is there anything you'd like to let me in on, Adam? I'd like to have something to laugh about too.

ADAM REARDON
(stung, hurt)
Sweetie, do you really think I would be insensitive to your illness?

Eve is saddened by the fact that she made assumptions about Adam that are probably not true.

EVE REARDON
(disappointed in
herself)
I'm sorry, Adam. I'm just not in the mood for frivolity.

ADAM REARDON
(argumentative)
I'm being anything but frivolous.

EVE REARDON
Then what are you being?

ADAM REARDON
I'm being hopeful.

EVE REARDON
You must know something I don't.

ADAM REARDON
(confident)
I know lots you don't know.

                    EVE REARDON
For instance...?

                    ADAM REARDON
Ouch.

                    EVE REARDON
I'm so sorry, dear. It's just that...

Her voice trails off; she is at a loss for words.

A light bulb goes off in her head

                    EVE REARDON (con't)
Does this have anything to do with
that delivery a few days ago?

                    ADAM REARDON
You're getting ahead of me.

                    EVE REARDON
Sorry.

                    ADAM REARDON
While you were focused on getting
rid of the creatures, I found myself going down some worm holes.
Would you like to hear what I
found?

                    EVE REARDON
          (half-heartedly)
Sure...yes.

                    ADAM REARDON
A German doctor named Otto Warburg
was awarded the Nobel Prize in 1931
for his research proving that cancer
cells use a form of non-oxygen
metabolism to survive. Cancer cells
are not like normal healthy cells.
The way they metabolize and create
energy for living and multiplication
is unique.

ADAM REARDON (con't)
Healthy cells are aerobic, meaning
that they function properly in the
presence of sufficient oxygen.
Healthy cells burn oxygen and
glucose, blood sugar, to produce
adenosine triphosphate OR "ATP",
which is the energy "currency" of
the cells. Cancer cells on the other
hand are anaerobic meaning they
function without oxygen. In the
absence of oxygen the cell reverts
to a primitive nutritional program
to sustain itself, converting
glucose, by fermentation.

Eve begins to get the drift of Adam's arguments. As
he progresses, she moves closer to the edge of the
sofa.

ADAM REARDON (con't)
The lactic acid produced by
fermentation lowers the cell pH:
acid/alkaline balance, and destroys
the ability of DNA and RNA to control
cell division...the cancer cells begin
to multiply unchecked. Warburg
emphasized that you can't make a cell
ferment unless a lack of oxygen is
involved. In 1955, two American
scientists confirmed Warburg's
findings. They found that oxygen
deficiency is ALWAYS present
when cancer develops. What Warburg
found, however, is you can reverse
fermentation simply by adding oxygen.
When you flood the cancer cell with
oxygen, you regain apoptosis, their
programmable cell death. If you put
enough oxygen into a cancer cell it
will turn on the Krebs Cycle: the
mitochondria and this reignites the
program for cell death. The therapies
that help deliver this much needed
oxygen include...

EVE REARDON

...a hyperbaric chamber...a million
dollar hyperbaric chamber.

ADAM REARDON

A minor impediment.

Eve Reardon is in limbo between hopeful but also
resigned.

EVE REARDON

What are you suggesting, Adam?

Eve is a step or two behind Adam's thinking.
He takes her hand and guides her into the
spare bedroom. There, a hyperbaric chamber
has been set up. Eve is in tears.

EVE REARDON

What did you have to do to get
this...sell your soul to the
devil?

ADAM REARDON

No. It was a gift.

EVE REARDON

We don't know any millionaires,
Adam.

ADAM REARDON

No, but we do know Major Ryan
Fitzpatrick.

EVE REARDON

But he works for the government,
he doesn't run it.

ADAM REARDON

Apparently, his bosses were very
appreciative of the work you've
done.

EVE REARDON

Then it's real?

She looks at the chamber, then to Adam. The
tears begin to flow. She is starting to
believe she might live...that she will live.